Natural Designs

THE EARLY MODERN AMERICAS

Series Editor
Peter C. Mancall

Volumes in the series explore neglected aspects of early modern history in the western hemisphere. Interdisciplinary in character, and with a special emphasis on the Atlantic World from 1450 to 1850, the series is published in partnership with the USC-Huntington Early Modern Studies Institute.

NATURAL DESIGNS

Gonzalo Fernández de Oviedo and the Invention of New World Nature

Elizabeth Gansen

PENN

UNIVERSITY OF PENNSYLVANIA PRESS

PHILADELPHIA

Published by
University of Pennsylvania Press
Philadelphia, Pennsylvania 19104-4112
www.upenn.edu/pennpress

Printed in the United States of America on acid-free paper
10 9 8 7 6 5 4 3 2 1

Hardcover ISBN: 978-1-5128-2456-8
eBook ISBN: 978-1-5128-2455-1

A Cataloging-in-Publication record is available from the Library of Congress

CONTENTS

Figure 1. Anteater. *Historia general*, 320r [bk. 12, chap. 21]. Private collection.

Figure 2. Anteater. *Historia general*, 321r [bk. 12, chap. 21]. Private collection.

PREFACE

Two little-known manuscript illustrations attest to the Spanish historian Gonzalo Fernández de Oviedo's lifelong efforts to visually capture New World nature. Located in the manuscript of *Book 12: On Animals* in the *Historia general y natural de las Indias* (1535–1549), the first depicts an anteater framed by two small trees, standing in front of a large mound of dirt (Figure 1). Tongue extended, the animal licks the compacted earth in search of its next meal. The image is remarkable not for what it portrays but for the multiple black crisscross lines that mar its surface. That these marks were made by Oviedo's hand is evident in the note he includes to the printer: "This figure must be removed from here, putting [in its place] that loose [drawing] with this signal ☆."[1] The "loose drawing" to which he refers is none other than an exquisitely rendered drawing of an anteater, which appears on an otherwise blank folio (Figure 2).[2] The interplay between the rejected drawing and its replacement is unprecedented in Oviedo's work. Certainly, Oviedo frequently "updates" the woodcuts from his earlier printed works in the later hand-drawn illustrations of the same in the manuscript of the *Historia general*. But readers are not privy to these artistic decisions in the same way that we are confronted with his choice here, to discard one image in favor of another.

Modern readers of Oviedo's magisterial *Historia general* would be hard-pressed to come across either of these illustrations. The manuscript of *Book 12* disappeared at the beginning of the twentieth century, reappearing briefly in 2016 when the William Reese Company, a bookseller in New Haven, Connecticut, sold it to an anonymous buyer, in whose possession it presumably remains. Before it was lost from view over a century ago, at least two scholars had consulted the manuscript—Juan Bautista Muñoz (1745–1799) and José Amador de los Ríos (1816–1878)—but neither mentions the crossed-out drawing. Following Oviedo's cues to the printer, in his transcription of the work

Muñoz copied the second illustration in place of the original. A century later, in the first modern edition of the *Historia general*, published in the 1850s, Ríos replaced all of Oviedo's hand-drawn illustrations with new ones that do not resemble their sixteenth-century predecessors.

Viewed side by side, these two drawings reveal the degree to which Oviedo continued to revise his illustrations of New World phenomena even in the final stage of his natural history project. It is clear that Oviedo found the second sketch of the anteater an improvement over the drawing it replaces. Its decidedly more lifelike nature—with its bushy tail, elongated snout, and long, thin tongue—accounts for his preference. Indeed, the stylistic differences between this drawing and the others in *Book 12*, coupled with the fact that the former was executed on a loose folio and later inserted into the manuscript, suggest that it was created by an unknown hand.[3]

What can we conclude from the existence of these two drawings, as well as the shadowy presence of another artist in the creation of the second? Although purely conjecture, I do not believe they are indicative of any far-reaching plan on Oviedo's part to replace the illustrations original to the manuscript of the *Historia general* with others, more professionally rendered—there is simply not enough evidence to support such a theory. Instead, I suggest that Oviedo was likely given the drawing by someone—perhaps an amateur artist who dabbled with sketching in his free time, much as Oviedo did—who knew of Oviedo's life work and either wanted to contribute with this single illustration or was asked to do so. In another one of Oviedo's writings, *Batallas y Quinquagenas*, he comments on having been handed scraps of paper on which were sketched the coats of arms and devices of some of Spain's leading citizens.[4] Ostensibly, Oviedo had made known his intentions to write a work on Spanish genealogy and heraldry; the individuals who provided him with this information contributed to those efforts. The drawing of the anteater strikes me as another instance in which such an informal exchange could have taken place. While nothing is known of this anonymous artist, it is likely that they resided in the New World and therefore based the sketch of the anteater on having observed it in the wild.

The *Historia general* cemented Oviedo's legacy as one of the foremost authorities on the New World in the early modern period. Yet the Spanish historian was only too aware of the enormous challenges of his task; several times during his career, he calls on others to assume the shared responsibility of recounting these events. The illustration of the anteater is symbolic of Oviedo's ultimate vision of the history of the Indies as a shared one. For as much as Oviedo continues to develop his artistic vision over time, his talents can take him only so far. By including this image of an anteater, executed by a foreign hand, Oviedo implicitly acknowledges the limits of his own artistic efforts. In doing so, he enacts New World natural history as the collective enterprise he knows it to be.

INTRODUCTION

The First Natural Historian of the Indies

In April 1493, a fifteen-year-old Gonzalo Fernández de Oviedo (1478–1557) looked on as a triumphant Christopher Columbus regaled Queen Isabella and King Ferdinand with news from his voyage in search of a westerly route to India and the Spice Islands. The Genovese explorer called the lands he had encountered the Indies, in reference to the famed geographies of the Far East.[1] In time, however, they came to be regarded as a New World, named as such for their conspicuous absence from the Western canon of knowledge. The fragmented glimpses of these far-off places that Columbus paraded before the royal couple—the native Taínos, parrots, bits of gold, and indigenous artifacts—made a deep impression on Oviedo, who recalled the event decades later in the first edition of the *Historia general de las Indias* (*General History of the Indies*) (Seville, 1535).[2] Still, as an adolescent recently appointed as a *mozo de cámara* at the royal court, little could Oviedo have imagined that twenty years later, in 1513, he would find himself preparing for an overseas journey of his own.[3] And then, the reality: on June 30, 1514, Oviedo stepped on the shores of the Caribbean for the first time, his senses reeling from the cacophony of sounds, smells, and sights that greeted him. Perhaps, as he took in this lush scenery, his thoughts flitted back to that spring day in Barcelona, so many years before. How much these memories must have paled, suddenly, in the face of these vivid surroundings! Whatever his first impressions of these lands, no one was more aware than he of the difficulties of communicating these lived experiences to those back in Europe: for the next four decades, until his death in 1557, Oviedo would fight to overcome the almost insurmountable gap between the physical existence of this New World and his attempts to describe it.

Oviedo is considered one of the earliest and most influential Spanish historians of the New World. Moreover, he was the first to chronicle the plants and animals of the Indies, topics that figure prominently in *Dela natural hystoria de las Indias* (*Of the Natural History of the Indies*) (Toledo, 1526), popularly known as the *Sumario*, the aforementioned *Historia* (1535), and the later manuscript of the same name, which Oviedo substantially edited and expanded from the earlier print edition. The natural world provokes as much wonder and delight for Oviedo as it does terror and mystery, exemplifying the close-knit but fraught relationship Spanish colonists sustained with the environment. A source of nourishment and other materials necessary for life, New World nature could just as soon occasion death if one were ignorant of its properties; in the *Historia* (1535), Oviedo writes of the sap of a tree now commonly referred to as the *manzanilla de la muerte* (little apple of death), so toxic that colonists who sought repose in its shade woke up to find themselves permanently blind.[4]

These are not anecdotes recounted by a distant spectator but by one who has experienced his fair share of heart-wrenching moments. Oviedo recalls a particularly terrifying incident involving his children who, not knowing any better, ate large quantities of a fruit that induced them to vomit without end. For several days, Oviedo despaired that they would not live; when they recovered, he vowed to eradicate all such plants from his property.[5] Another episode on the island of Hispaniola convinced Oviedo that his own death was imminent: not realizing the red tint of his urine was a consequence of having consumed the prickly pear fruit, he concluded that all the veins in his body had burst and that his life's blood was draining out of him.[6] Oviedo lives these moments of great distress alongside others that occasion joy and wonderment: the pineapple prompts him to elevate his prose to near poetic registers while the spiny cactus utterly befuddles him. The encounters Oviedo sustains with these New World phenomena, and which he commemorates in his writings, are unparalleled among early modern accounts of the Americas. His championing of the flora and fauna of the Indies as central to the events of the conquest and colonization marks him as the founder of Latin American natural history.

* * *

Natural Designs foregrounds Oviedo's role as a writer, illustrator, and editor of New World nature. The Spanish historian traveled to the Caribbean for the first time in 1513, at the age of thirty-five, and he published his initial impressions of its natural world in the *Sumario* (1526), at the age of forty-eight.

Oviedo continued to revise and augment these descriptions of flora and fauna during the next several decades, publishing the first part of the *Historia general* in 1535 and entertaining frustrated hopes in the 1540s for the printing of the complete work (a revised and expanded part 1, in addition to parts 2 and 3, which are conserved in the Monserrat manuscript). His writings, I argue, cannot be properly understood without an awareness of their chronological development over the periods of Oviedo's long life (1478–1557) in Spain and the Americas. Thus, in these pages I undertake a systematic, comprehensive treatment of his approach to New World plants and animals. As Oviedo strove to represent the natural world of the Indies in word and image, he faced challenges and dilemmas that others had never had to consider. I trace these developments from their initial appearance in the *Sumario* to their final form in the Monserrat manuscript (1535–1549).

In exploring Oviedo's natural history project as a product of the Old World and the New, this book bridges two areas of inquiry that have been the subject of numerous studies of late. The first explores natural history as it began to take form in Europe in the late fifteenth century, emerging definitively as a discipline in the 1530s and 1540s. Scholars such as Brian Ogilvie, Alix Cooper, and Paula Findlen have explored the growing empirical culture of observation, recording, and illustration that came to define the work of these early modern natural historians.[7] They demonstrate how this literary genre drew inspiration as much from local natural environments as from far-off locales, as aficionados from a range of professions and social classes began to explore the physical world surrounding them. These individuals struck up friendships with others who shared their interest, expressing their enthusiasm for the natural world via epistolary exchange and through the gifting of specimens and seeds. In this book, I aim to draw attention to Oviedo as a product of and participant in these cultural and scientific developments. He learned as much from the humanists with whom he came in contact in Renaissance Italy (1499–1502) as they learned from him, communicating with the likes of Giovanni Battista Ramusio (1485–1557) and Pietro Bembo (1470–1547) in subsequent decades about his American experiences.

The second area of inquiry to which this book contributes is Europe's encounters with New World nature and, more specifically, efforts on the part of Spanish colonists, religious institutions, and representatives of the Crown to learn about and document the vast biological holdings of Spain's overseas empire. Scholars including Antonio Barrera-Osorio, Arndt Brendecke, and

Daniela Bleichmar have traced different manifestations of these initiatives. During Oviedo's lifetime, the study of nature was often proposed and undertaken by individuals—merchants, entrepreneurs, and artisans—with a view toward personal gain.[8] It was only gradually that many of these practices became institutionalized, as the Spanish Crown sought to bring them under its purview as instruments of empire.[9] By the end of the sixteenth century, New World nature had become the object of a government-sponsored scientific expedition under the direction of the court physician Francisco Hernández (ca. 1514–1587). It was only with the botanical voyages of the late eighteenth and early nineteenth centuries, however, that royal interest in the economic possibilities of these resources reached its height.[10] As scholars such as Jorge Cañizares Esguerra and Andrés I. Prieto have demonstrated, the Spanish Crown was not the only participant in this burgeoning awareness of nature's potential; New World natural phenomena were also enlisted as tools of evangelization and, in later centuries, as a means of asserting Creole identities.[11]

The narrative of New World nature's growing importance at the hands of various institutions and causes is well known. Often, however, the status of the natural world in the initial decades of Spain's conquest and colonization of the Indies is afforded only glancing treatment.[12] As much as Oviedo is credited as a pioneer in studies of American nature, his contributions to early modern conceptions of the flora and fauna of the Indies are still not widely understood nor appreciated. Yet Oviedo's influence transcends his own lifetime: the *Sumario* and the *Historia general* offered a model for subsequent natural histories, and they served as indispensable references for future historians, scientists, and travelers.[13] *Natural Designs* returns to the early sixteenth century, to reappraise the works of he who first gave shape to New World nature.

Although increasing attention has been devoted to Oviedo in his capacity as a natural historian, his writings on this subject tend to be overshadowed by the larger scope of the project he sought to undertake. Oviedo's ultimate objective of writing a history that embraced the natural and human world means that these two dimensions of his task are often treated simultaneously in scholarship. Antonello Gerbi's now classic *Nature in the New World: From Christopher Columbus to Gonzalo Fernández de Oviedo* (1975) epitomizes this tendency. The Italian historian's approach to Oviedo's life and writings is encyclopedic and wide-reaching, broaching such diverse themes as his medievalism, his account of the American tiger, and his portrayal of the Mexican conquistador Hernando Cortés. In *Fernández de Oviedo's Chronicle of America: A New History for a New World* (2007), Kathleen Ann Myers similarly considers subjects pertaining to both natural and human history, respectively. These studies provide a sense of the totality of Oviedo's written output on the Indies and the themes and concerns that resonate throughout them. While the merits of this method are undeniable, it is necessarily limited by the vast breadth of material covered, obscuring the particularities of Oviedo's experiences as a natural historian.

As much as Oviedo understood New World nature as inextricably connected to the events of human history, the former occupies a special place in his writings. In the *Sumario*, natural history materializes as a discrete genre, independent of its human counterpart. Here, for the first time, Oviedo subjects Caribbean plants and animals to systematic treatment—classifying, organizing, and describing these phenomena and their utility for humankind. In the *Sumario*, as well as in the *Historia* (1535) and the Monserrat manuscript, the literary construction of nature operates according to criteria inherently different from those which drive the narrative of Spanish exploits in the Americas. In this book, I take as my premise that to truly understand Oviedo's task as a natural historian, his account of New World nature needs to be considered separately from the larger narrative of which it forms a part. From this more focused analysis emerge the concerns and strategies that are the hallmarks of Oviedo's contributions to the literary genre of natural history.

I build on the work of other scholars focused primarily on Oviedo as a natural historian.[14] Enrique Álvarez López (1897–1961) authored several lengthy articles on Oviedo's treatment of New World flora and fauna.[15] A natural historian in his own right, Álvarez López's interest in Oviedo's writings is primarily tied to the biological novelties he documents. Neither Álvarez López nor Gerbi (1904–1976), his successor by several decades, discusses the illustrations of plants and animals—printed as woodcuts, in the case of the *Sumario* and the *Historia* (1535) and hand-drawn in the Monserrat manuscript—that Oviedo considered vital to his account of the natural history of the Indies. That their regard for Oviedo's work does not extend to its visual material indicates more about their judgment on his artistry than the function of illustration in sixteenth-century natural history.

Of the more recent scholars to have taken up this topic, Jesús Carrillo Castillo's contributions stand out for their academic rigor and analytic depth. Above all, his monograph, *Naturaleza e imperio: La representación del mundo natural en la "Historia general y natural de las Indias" de Gonzalo Fernández de Oviedo* (2004), provides much-needed insight into Oviedo's role as a natural historian of the Indies. Carrillo Castillo explores the courtly culture of turn-of-the-sixteenth-century Spain and Italy with the purpose of shedding light on the humanistic environment in which Oviedo was immersed for much of his early life. He pays particular attention to the Monserrat manuscript, analyzing its development as a continual work in progress. Carrillo Castillo also highlights the illustrations in the *Sumario* and the *Historia general* and the changes they undergo over time and in the hands of various translators and editors. I continue this work, considering Oviedo's natural history at large as a decades-long process of writing and revising both word and image.

As Carrillo Castillo demonstrates, it is not sufficient to consider Oviedo's writings on the Indies in aggregate. Too often, however, scholars refer to the *Historia general* without distinguishing between its different forms and phases.[16] As a result, the *Historia* (1535) is frequently subsumed into the later Monserrat manuscript (1535–1549). Although related to one another, the two are not the same; Oviedo expands on and edits the former in the latter, in addition to including two new parts. Yet scholars regularly treat them as one or inadvertently attribute statements that Oviedo makes in the Monserrat manuscript to the *Historia* (1535).[17] Viewing the *Historia general* as a single work gives the false impression that Oviedo maintained a static, unchanging outlook on New World nature. In contrast, I consider each work as a discrete contribution that Oviedo makes to the study of this subject. Written decades apart from one another under varying circumstances and centering on different geographies, the *Sumario*, the *Historia* (1535), and the Monserrat manuscript correspond to different stages in Oviedo's constantly evolving project. It is only by studying each individually that we can attain a more nuanced perspective on Oviedo's task as a natural historian of the Indies.

To trace the origins and later development of Oviedo's writings on natural history most effectively, I have organized the chapters chronologically. Chapters 1 and 2 address Oviedo's intellectual formation during his early life in Spain and Italy (1478–1513). I argue that the time he spent at the Castilian royal court (1490–1498, 1503–1512) and in Italy (1499–1502), played a pivotal role in how Oviedo conceived of the natural world at large. Pliny the Elder's (23–79 CE) *Historia naturalis* was much-discussed and read in these aristocratic circles; eventually, Oviedo would use the Roman historian's natural history as a model for his own, albeit with an acute awareness of the unprecedented situation that he faced in writing about a part of the world not dealt with by classical authors. Nature in early modern Europe was not just something to be read about but also experienced. The burgeoning interest in Italy regarding nature in all its forms—as a product of its natural environment, cultivated in gardens, and represented in paintings—awaken Oviedo to its importance. The knowledge that he gained in these courtly and humanistic milieus would prove decisive to his encounter with the flora and fauna of the Indies.

Chapters 3 and 4 explore Oviedo's initial incursions into New World nature in the *Sumario* and the *Historia* (1535), respectively, and how they respond to certain ideological and practical considerations and needs. An important through line of the book, which emerges here, concerns the editorial choices Oviedo makes in his attempts to impose order on what for him is a vibrant and virtually unknown natural world. In the *Sumario*, for the first time, Oviedo faces a dilemma that would resurface in his subsequent writings regarding how to best organize his account of New World flora and fauna. A fundamental consideration for Oviedo when making these decisions is the varying circumstances of his life in the Caribbean, and particularly his long years on the island of Hispaniola.

Significantly, Oviedo was not just a writer of New World nature but an illustrator of its wonders. The four woodcuts in the *Sumario* are the beginnings of a visual repertoire that gave pictorial form to the likes of the pineapple, the iguana, and cacti, representing the first images of their kind to circulate in Europe. Oviedo's career is marked by an increasing awareness of the limitations of the written word to convey reality in all its forms, as well as struggles with how best to depict flora and fauna visually. His decision to include visual representation in his writings on the Indies took place contemporaneously with the publication of the first illustrated natural histories in Europe, Otto Brunfels's *Herbarum vivae eicones* (*Living Plant Images*) (1530) and Leonhart Fuchs's *De historia stirpium commentarii insignes* (*Notable Commentaries on the History of Plants*) (1542). In Chapters 5 and 6, I chart the creation and evolution of visual representation in Oviedo's writings, and specifically of the woodcuts and hand-drawn images of plants, connecting them to these concurrent developments in early modern Europe.

Finally, in Chapter 7, I consider the final iteration of Oviedo's natural history project, the Monserrat manuscript (1535–1549). A palimpsest text, the manuscript reveals the process by which Oviedo added, deleted, and shaped the "final" version of his *Historia general*. Originally, Oviedo had determined to divide New World nature between the work's three parts. At some point, however, he found this structure wanting and decided to move all this information to part 1.[18] I maintain that Oviedo undertook these editorial revisions in the 1530s as a response to some of the same factors informing the organization of nature in his earlier published works, the *Sumario* and the *Historia* (1535).

Oviedo's writings on the Indies cannot be understood in isolation from the facts of his life in early modern Europe and the Americas. To that end, the section that follows offers a brief biographical sketch of Oviedo. By no means exhaustive, nevertheless it outlines the major periods of his life and traces the progression of his historiographical career. I will explore some of these episodes in more detail in later chapters.

The Life and Writings of Gonzalo Fernández de Oviedo (1478–1557)

From a young age, Oviedo was exposed to the courtly culture that constituted the basis of his education, and which later informed his approach to New World nature.[19] At some point in the 1480s, Oviedo entered the service of the Duke of Villahermosa, Alfonso de Aragón y de Sotomayor (1479–1513). Thanks to the Sotomayors' connections to the royal family—the young duke was King Ferdinand's nephew—by 1490 Oviedo had begun to accompany the Catholic Monarchs' itinerant court as a *mozo de cámara* in the retinue of their son Prince John (1478–1497). Oviedo joined the court just in time to see fulfilled the centuries-old ambition of Spain's Christian kingdoms to wrest control of the peninsula from the Muslims, a moment that coincided with the climax of the monarchs' political power. In 1492 Oviedo witnessed the conquest of the last Moorish kingdom of Granada as well as the signing of the Capitulations of Santa Fe, which set forth the conditions under which Columbus would undertake his inaugural voyage. Though an adolescent, Oviedo seems to have grasped the historic import of these events, which he recorded in a diary.[20]

On April 2, 1497, Prince John married Princess Margaret of Austria, daughter of Emperor Maximilian of Austria. Yet wedding bells quickly gave way to tears and laments: a mere six months after the festivities, the prince died. With his world suddenly upended, Oviedo made the decision to leave Spain: "In this way, there [with the prince's death] ended the hopes that I sustained in what this mortal life had to offer me. My discontentment led me to leave Spain and wander through the world."[21]

For the next four years, from 1499 to 1502, Oviedo traveled through Italy. The experiences he had there were formative for his humanistic education, and he alludes to them time and again in his writings. He spent four and a half months in Genoa in 1499, where he served one of its governors, Giovanni Adorno.[22] From 1488 to 1499, Genoa was held by the Duke of Milan Ludovico Sforza (1452–1508) as part of the Duchy of Milan. Oviedo benefited greatly from the duke's patronage of the arts: it was at his court that Oviedo claims to have met Leonardo da Vinci, who had been in Ludovico's employ since 1489 and with whom he conversed about the pictorial arts.[23]

Sometime before Louis XII's impending invasion of Milan and Ludovico's withdrawal to Tirol in September 1499, Oviedo left Milan and journeyed to the court of Isabella d'Este, the Marchesa of Mantua and Ludovico's sister-in-law (her younger sister, Ludovico's wife, Beatrice d'Este, had died in 1497). Even more so than her brother-in-law, Isabella was a supporter of the arts; Oviedo's acquaintance with the Venetian painter Andrea Mantegna (1431–1506) likely occurred in her court, as the artist was a longtime painter for her husband's family in Mantua. It is not clear how long Oviedo remained in Isabella's household before entering the service of Cardinal Juan de Borgia, but it could not have been more than a handful of months. Unfortunately, Oviedo's employment to him ended almost as soon as it had begun; after several months, the cardinal died suddenly on January 17, 1500.[24]

By April 1500, after this whirlwind of travel from Milan to Mantua, Turin, Pavia, Bologna, and Rome (the latter four cities as part of the entourage of the cardinal), Oviedo arrived at what would be his final destination in Italy—Naples. Here, he served as an *ayudante de cámara* or valet to King Frederick (1452–1504, reigned 1496–1501).[25] Oviedo admired the king's gardens in the villa of Poggio reale and maintained that he surpassed all other princes in the "richness of his chambers, royal chapel, hunting and stables."[26]

In Naples, Oviedo became acquainted with several of the most prominent poets and humanists of the period. He attended meetings of the Accademia Pontaniana, whose members included its founder, Giovanni Gioviano Pontano (1426–1503), and the poet Jacobo Sannazaro (1458–1530). Pontano had retired several years earlier, in 1498, from a life of politics (he spent part of the 1490s as first secretary to King Alfonso II of Naples, and later to Ferdinand II).[27] Oviedo's residence in Naples coincided with some of the most prolific years in Pontano's writing career: in 1498, he celebrated the publication of five treatises he had written on the moral obligations of the wealthy, and in 1499 he wrote *Actius*, considered the first Renaissance treatise on the art of writing history (*ars historia*).[28] Oviedo greatly respected and admired Pontano, whom he recalls having heard speak on several occasions.[29]

Oviedo's time among the members of the Accademia Pontaniana concluded prematurely when vying French and Spanish interests put an end to King Frederick's rule of the kingdom of Naples. Unbeknownst to the monarch, on November 11, 1500, King Ferdinand and Louis XII had signed the Treaty of Granada in which they divided the kingdom between themselves, alleging that King Frederick had betrayed the Christian faith by consorting with the Turks.[30] Recognizing that King Ferdinand was motivated by his desire to protect Naples from the French, Oviedo does not openly criticize his role in his cousin's downfall. Still, Oviedo defends King Frederick from the accusations made against him and calls him "a most excellent prince."[31] Anticipating the impending invasion of Naples, King Frederick arranged for his family to abandon the kingdom; he sent his son, the Duke of Calabria, to Taranto, Italy, and he requested that Oviedo accompany his half-sister Joanna of Naples (1479–1518) to Valencia as her *guardarropa*.[32] Once his services to the queen had terminated, Oviedo journeyed to Zaragoza and later to Madrid, where he would reside for the next twelve years, his travels coming to a momentary end.

Although Oviedo never returned to Italy, the years he spent there had lasting consequences for his life and writings. Most immediately, the Catholic Monarchs looked favorably on his connections with Italian royalty, which allowed him to continue to benefit from the privileges of a life at the royal court: in 1503, shortly after Oviedo returned to Spain, King Ferdinand requested that he serve the Duke of Calabria, son of the deposed King Frederick of Naples.[33] There is no indication in Oviedo's writings of when his services to the duke ended, but his references to other commissions undertaken for King Ferdinand support the idea that he continued to move in close proximity to the royal court for at least the next several years. Oviedo recalls that at the Courts of Toro (*Cortes de Toro*) in January 1505, which King Ferdinand had convoked to solicit support for his regency of Castile after Queen Isabella's death, the monarch had encouraged him to write a history of Spain.[34] Oviedo would not complete this project, the *Catálogo real*, until 1532, and it remained unpublished upon his death in 1557.

By 1512 Oviedo had become the secretary of Gonzálo Fernández de Córdoba (1453–1515).[35] As such, he planned to accompany the *Gran Capitán*, as he was known, on a military expedition to Italy against the French; like many others, Oviedo invested financially in its preparations. However, before the troops could depart, Fernández de Córdoba received orders from King Ferdinand to journey on alone, as soldiers would be provided for him upon his arrival in Italy. Through a single stroke of misfortune, Oviedo found his plans for the future foiled yet again. The efforts on the part of Oviedo and his comrades to recoup their capital met with mixed results, and many of them began to look for other avenues for profit. Several, including Oviedo, saw potential wealth in the expedition of Pedrarias Dávila (1440–1531) to Castilla del Oro, in the Indies. Suddenly, at the age of thirty-five, Oviedo found himself contemplating a life overseas.

On July 27, 1513, King Ferdinand appointed Pedrarias Dávila to the governorship of the recently created province of Castilla del Oro, a mainland territory on the Caribbean coast of southern Central America and northwestern South America.[36] Initially, Oviedo was appointed to the expedition as *escribano mayor de minas, del crimen, y juzgado del herrar los indios*.[37] When the *veedor de minas y fundiciones*, Juan Quincedo, died before the expedition set sail, Oviedo assumed his duties as well.[38] In the more than four decades that he resided in the Indies, Oviedo would serve the Crown in various administrative capacities, just as he had in Spain. More importantly, his years overseas would also provide him with the material to become a historian, an aspiration that Oviedo appears to have had as far back as his adolescence at the Castilian royal court, and which had begun to take shape in 1505, with King Ferdinand's request that he write a history of Spain.

The expedition of two thousand men and twenty-two ships arrived at the port of Santa María, in the province of Castilla del Oro, on June 30, 1514.[39] No sooner had Oviedo disembarked on the shores of the mainland than he witnessed, aghast, what he referred to as "the beginning of the destruction of the land (that they [the conquistadors] refer to as 'pacification and conquest')."[40] He attributed these crimes to Pedrarias's corrupt leadership; writing of these events years later in the *Historia general*, with the wisdom afforded by hindsight, Oviedo denounces the conquistador's actions as having set the precedent for other, similar crimes in the Indies.[41] Within six months of his arrival in Castilla del Oro, Oviedo decided to return to Spain and request an audience with King Ferdinand to voice his concerns about this ongoing situation and advocate for Pedrarias's removal as governor.

By the time Oviedo disembarked in Seville in December 1515, King Ferdinand's health was failing. Oviedo met with him briefly in Plasencia, but the king made it clear that a formal audience would not take place until he had completed his journey to Seville.[42] This was the last time Oviedo would see him: on January 23, 1516, King Ferdinand died. Upon learning of his death, Oviedo traveled to Flanders, Belgium, to pursue his agenda with King Ferdinand's grandson and heir, the newly crowned King Charles I of Castile (and soon-to-be Holy Roman Emperor Charles V). Since the Royal Council of Castile (*Consejo real de Castilla*) was little equipped to deal with New World affairs, the king decided to entrust these matters to the recently appointed dual regents of Castile, Cardinals Francisco Jiménez de Cisneros (1436–1517) and Adrian of Utrech (1459–1523).[43] A discouraged Oviedo returned to Spain to submit his report to the two cardinals, but their attention was focused elsewhere and his appeals did not have the desired effect.

Despite these setbacks, ultimately Oviedo achieved his objective: in 1519, the monarch and newly elected Holy Roman emperor appointed Lope de Sosa as governor of Castilla del Oro. In the intervening years, while Oviedo waited for a royal audience, it is likely that he composed the chivalric romance *Libro del muy esforçado e invencible cauallero de la Fortuna propiamente llamado don Claribalte*, which was published in Valencia in 1519. He dedicated the work to the Duke of Calabria, who had been imprisoned since 1512 in Játira, a fortress near Valencia, after an attempt to reunite with his family in France. His writing of *Claribalte* is somewhat surprising, given his condemnation of the genre in the *Historia general*.[44] As a former servant to the duke, and one whose plight he sympathized with, perhaps Oviedo was catering to the duke's reading habits despite his aversion to this type of literature.

Satisfied with what he had managed to accomplish in Spain, Oviedo departed for the Indies sometime at the beginning of 1520. Yet when he arrived in Castilla del Oro, he learned that Sosa had died even before setting foot in the territory he had been assigned to govern.[45] His death meant that Pedrarias continued to rule, with more impunity than ever. Three years later, unable to tolerate the hostility of Pedrarias's allies toward him as well as the deteriorating conditions of life in Castilla del Oro, Oviedo embarked for Spain again. Arriving there on November 5, 1523, he remained at court for over two years, continuing to advocate for Pedrarias's removal as governor.[46] Finally, on August 25, 1525, the Council of the Indies named Pedro de los Ríos to take Pedrarias's place.

Oviedo took advantage of his second stay in Spain to further his personal ambitions. In Toledo on February 15, 1526, Ramón de Petras printed *Dela natural hystoria delas Indias*, or the *Sumario*. According to Oviedo, the work emerged from his desire to introduce Charles V to the vibrant natural world of the monarch's overseas territories.[47] Evidently, the emperor recognized the importance of such an endeavor, and ordered Oviedo to print it (at his own expense) so as to make it available for others.[48] Not only is the *Sumario* the first writing on the Indies to focus exclusively on New World nature but it is also the first to depict visually aspects of this reality. Its four woodcuts—including a tree of an unknown species, a hammock, a plantain leaf, and sticks used for making fire—constitute some of the first illustrated material from the Caribbean to circulate in Europe. Following the publication of the *Sumario*, Oviedo embarked for Panama and then Nicaragua, where he remained from 1527 to 1529.

The subsequent decade marks a watershed period in Oviedo's career as a historian of the Indies, which began upon his departure to Spain, sometime in March 1531.[49] As always, Oviedo had a mixture of personal and professional motives for his journey, most important among them being his pursuit of an official appointment as chronicler of the Indies (*cronista de Indias*). The municipal council of Santo Domingo wrote to Charles V in March 1531 (probably at Oviedo's behest), urging the monarch to award Oviedo the title and salary befitting such a position so that he could continue his "General Chronicle and Natural History of these Indies."[50] As if to lend credence to the council's enthusiastic support of his literary pretensions, at the end of April 1532, Oviedo presented his *Catálogo real* (1505–1532) to Queen Isabella of Portugal (1503–1539). In its preface, Oviedo describes the tome as a synthesis of the history of Spain and its monarchs, taken from a variety of sources and written at the direction of Charles V's deceased grandfather, King Ferdinand.[51] Although Charles V was not available to receive the manuscript, as Oviedo had hoped, nevertheless he offered it to the queen with the desire that it be useful in her son Prince Philip's education.[52]

Shortly after Oviedo's audience with Queen Isabella, on May 7, 1532, the Council of the Indies composed a letter in support of the municipal council's request to appoint Oviedo as *cronista de Indias*.[53] Several months later, on August 18, 1532, Charles V approved the position, granting Oviedo a salary of 30,000 maravedís and the authority to request documents from Spanish officials anywhere in the Indies.[54] Oviedo accepted the post and, at the same time, renounced his title as *veedor de minas y fundiciones*, an office he had occupied for eighteen years, from 1514 to 1532.

No sooner had Oviedo arrived in Santo Domingo than he embarked again for Spain. A letter addressed to Charles V from the Royal Audience of Santo

Domingo, dated April 27, 1534, states that Oviedo had been sent as *procurador*, charged with addressing the economic, administrative, and political problems confronting Hispaniola.[55] In his writings, Oviedo does not remark on these bureaucratic undertakings. Instead, his account of this fourth trip to Spain highlights his personal and literary accomplishments. To begin, on September 30, 1535, Juan Cromberger published in Seville the first of three parts of Oviedo's *Historia general*, which details the human and natural histories of the Caribbean islands of Hispaniola, Cuba, San Juan (Puerto Rico), Jamaica, Cubagua, and Margarita. A testament to the contemporaneous success of the work are its early translations into Italian and French: reprinted in 1547, it was translated in full into Italian in 1556 in Venice and partially into French in 1555 and 1556 in Paris. In the fourteen years after the publication of the *Historia* (1535), Oviedo substantially edited and expanded on the work. The product of these efforts, contained in the Monserrat manuscript, consists of Oviedo's revisions to the published part 1 as well as an additional two parts, relative to the geographies of mainland South and Central America and parts of North America.

Shortly before Oviedo's return trip to Santo Domingo, in April 1535, he was asked to provide Juan de Zúñiga y Avellaneda—*comendador mayor* of Castile and tutor to Prince Philip—with the details of the appointments and responsibilities of those who had served Prince John.[56] Oviedo states that this request came at the behest of Charles V, who was ostensibly aware of the importance the late Queen Isabella had placed on the education of her son and wished to follow her example. Given his impending voyage, Oviedo could not spend as much time with the commission as he would have liked, but he was able to dispatch a short, five-page document to Zúñiga before his departure. Later, during his final voyage to Spain in 1546, Oviedo wrote a more complete account of the organization of Prince John's household, titled *Libro de la cámara real*.

Upon Oviedo's return to Santo Domingo in 1535, there seems to have been a permanence and uniformity to his life that had not previously existed: until his death, in 1557, Oviedo remained on the island, dedicating himself to his twofold responsibilities as *alcaide*, a position to which he had been appointed in 1533, and *cronista de Indias*.[57] He interrupted his domesticity only once, from 1546 to 1549, to travel to Spain as *procurador* of Santo Domingo.[58] After an audience in Madrid with Prince Philip, who was governing in his father's absence, Oviedo traveled to Salamanca.[59] It was there that he likely commissioned Juan de Junta to print the second edition of the first part of the *Historia*, titled *Crónica delas Indias: La hystoria de las Indias agora nuevamente impressa corregida y emendada.*[60] This was the last of Oviedo's writings that he personally saw in print in Castilian during his lifetime.

Oviedo returned to Santo Domingo in 1549 and continued to work on several literary projects: the *Historia general* and two other manuscripts, *Quinquagenas de la nobleza de España* and *Batallas y Quinquagenas*. These last two endeavors share a common theme—Spanish nobility—but differ substantially in their presentation. *Batallas*, which Oviedo began in 1535, explores "distinguished and noble lineages and famous nobleman" through a series of dialogues between two of Oviedo's literary personae: one, an elderly caretaker (Alcaide) of the fortress of Santo Domingo, and the other, a young protégé-like character, Sereno.[61] *Quinquagenas*, composed by January 10, 1555, consists of a series of verses of fifty stanzas of fifty verses each (which Oviedo called "segunda rima"), divided into three parts. In the "Notice of Intent" (*aviso de intento*) of *Quinquagenas*, Oviedo explains the twofold purpose of his work: to correct vice and praise virtue and to bring to mind the famous men of Spain.[62]

In the prologue of *Quinquagenas*, sent to Prince Philip probably sometime in 1555, Oviedo explains that the second and third parts of the *Historia general* were being published.[63] Ultimately, however, only *Book 20* was printed (Valladolid, 1557). The scholar Daymond Turner conjectures that Oviedo had initiated the publication of at least some of the second book, but that its printing was interrupted for lack of funds. When the printer Fernández de Córdoba heard news of Oviedo's death, he decided to salvage those sections of the work that had already come off the press—namely, *Book 20*.[64] The publication of parts 2 and 3 of the *Historia general* would have to wait until the nineteenth century, when José Amador de los Ríos published a complete edition of the work (1851–1855).

As complaints of declining health, deafness, and old age began to invade his writings, Oviedo took advantage of what little time remained to him to try to complete his various projects: rising before dawn, he spent his last years in the fortress of Santo Domingo reading and writing.[65] Despite Oviedo's best efforts, the *Historia general* and *Batallas* remained, to a greater or lesser extent, unfinished upon his death in 1557. While Oviedo would have lamented this fact, so driven was he to complete these writings, his life could not have ended any other way. His lifeline to the world was the manuscript in progress, the next chapter to be written, and the next project to be undertaken, with nothing, short of death, standing in his way.

* * *

Note on citation: To differentiate between the print edition of the first part of the *Historia general* and the later manuscript copy of the edited and expanded version of the work, I refer to the former as the *Historia* (1535) and the latter interchangeably as the *Historia general y natural de las Indias*, the *Historia general*, and the Monserrat manuscript.

CHAPTER 1

Nature on Display

Oviedo in the Early Renaissance

A thirty-six-year-old Gonzalo Fernández de Oviedo embarked on his inaugural voyage to the New World in 1514, taking with him memories that he would cherish for the rest of his life. By the time he boarded one of the twenty-two ships bound for Santa María del Antigua del Darién under the command of Pedrarias Dávila, Oviedo had experienced several decades as a courtier at some of the most renowned courts of early modern Spain and Italy. His service for the Castilian royal household in the 1490s and 1500s, as well as at the courts of Italian noblemen and royalty in Milan, Mantua, and Naples (1499–1502), afforded Oviedo the opportunity to interact with some of the most prominent figures of the period and to participate in the courtly culture that his patrons fostered. Oviedo's constant allusions to these experiences in his writings mark these years as the most intellectually formative of his life.

Oviedo's presence in these courtly milieus at the turn of the sixteenth century had a pivotal role in shaping his approach to the natural world of the Indies. There was no contemporary model on which he could base his treatment of this topic in *Dela natural hystoria delas Indias* (1526), commonly known as the *Sumario,* and the *Historia general y natural de las Indias.* Nevertheless, natural history was a subject of interest for many. A myriad of scholars, travelers, artists, physicians, princes, and patricians—some with whom Oviedo was acquainted—were involved in exploring and promoting the world around them, albeit to a large degree independently of one another. By the 1530s and 1540s, these varied activities had culminated in the emerging discipline of natural history: in Italy, university chairs were created for the teaching of medical simples—the study of individual plants used to produce remedies—and botanical gardens were established at the universities of Pisa and Padua. As Paula Findlen demonstrates, these accomplishments were not radical moments of discovery but rather the result of gradual shifts in perceptions of the natural world throughout the fourteenth and fifteenth centuries.[1] It is precisely in Oviedo's lifetime that the consequences of these changes became most apparent, as increasingly Oviedo and his contemporaries came to understand the study of the external, physical world as a necessary complement to book learning.

Much like other amateur natural historians applied their subject-specific expertise to making sense of their surroundings, Oviedo approached New World flora and fauna by drawing on his courtly education. Learned disagreements in the 1490s concerning the origin of the errors in the Roman historian Pliny the Elder's (23–79 CE) *Historia naturalis* highlighted the changing attitudes toward the study of nature in the aristocratic circles Oviedo frequented in Italy. In addition to discussing the merits of Pliny's work, participants in these literary debates increasingly began to uphold eyewitness testimony as an indispensable companion to classical knowledge of the natural world. The popularity of the works of the Roman agricultural writers as well as the agronomical treatises of Piero de' Crescenzi (1230/35–ca. 1320) and Gabriel Alonso de Herrera (1470–1539) gave rise to a vision of nature that promoted the idyllic virtues of country living, with its balanced routine of labor and leisure. Similar values were embodied in the urban and rural gardens of Italian and Spanish patricians and monarchs, at the same time as the variety and beauty of the plants displayed testified to the power and prestige of their owners. In short, nature assumed a variety of guises in the period: as an ornament of wealth, a literary and intellectual pursuit, and a source of work and rest. From these different yet complementary perspectives become visible a set of collective practices and beliefs that shaped the role of nature at large in early modern Spain and Italy. Necessarily, the gaze that Oviedo projects on the New World in the *Sumario* and the *Historia general* takes these earlier experiences as its point of departure.

Pliny the Elder's *Historia naturalis*: A Classical Text Reimagined

In the treatise *De tradendis disciplinis* (*On Education*), published in 1531, Oviedo's contemporary and compatriot, the Spanish scholar Juan Luis Vives

(1493–1540), endorses the benefits of the study of the natural world by all those desirous of learning. The curriculum he advises for such a pursuit is one in which the contemplation of nature supplements book learning, as the senses "open up the way to all knowledge."[2] He recommends the study of "outward nature by close observation," a task that requires that the eyes, ears, and mind become attuned to its innate characteristics and seasonal changes.[3] Although Aristotle too had maintained these practices as a basis for the study of natural causes, natural philosophers in the Middle Ages tended to reject this empirical approach to knowledge in favor of book learning. They sought answers to their queries regarding nature in authoritative texts, engaging in dialectical argumentation that produced no practical results and that was rooted in speculation rather than fact.[4] A successor of these medieval intellectuals, Vives understands observation of nature as but a preliminary step to what he conceives as a deeper, more complex engagement with the functioning of the natural world at large. Nevertheless, that he positions sensory experience as an essential precursor to knowledge acquired from written tradition points to the renewed respect it had attained since the Middle Ages.

Indeed, by the time Oviedo began writing about New World nature, in the 1520s and 1530s, natural history had already become a subject embraced by Europe's intellectual elite. A catalyst for its welcome reception by Vives and other scholars in the mid-sixteenth century was the controversy that arose in the 1490s concerning Pliny the Elder's *Historia naturalis*. The contours of this dispute were philological, its participants pitting wits over the source and number of errors they identified in the Roman historian's work. While the disagreement centered on the accuracy of the information Pliny provided, there was a growing awareness among those involved of the need to look beyond the venerable words of the ancients to the natural world around them. Oviedo's treatment of Pliny's magnum opus in the *Sumario* and the *Historia general* is a consequence of this gradual transition from a textually based approach to the classical corpus to one based on eyewitness accounts and personal experiences.

Unlike many Greek and Roman works, which were unknown in medieval Europe and rediscovered in the Renaissance, Pliny's *Historia naturalis* enjoyed popularity throughout the Middle Ages. Consequently, it was one of the first books to come off the printing press: an edition of the text was issued in 1469 by Johannes de Spira in Venice and several more editions were published in the 1470s.[5] The ability to create and circulate identical copies of a text made possible by the invention of movable type presented humanists of the late fifteenth century with difficulties never before faced. As they strove to produce critical editions of the works of classical authors, they became aware of the existence of hundreds of thousands of inconsistencies and errors in codices and extant print editions, the result of a centuries-old tradition of copying and recopying these texts. In the case of Pliny's *Historia naturalis*, this impression was particularly acute; as a synthesis of prior Greek and Roman scientific works, many of which had since been lost, humanists relied on Pliny to supply the Latin equivalents to the Greek words used by Aristotle, Dioscorides, and Theophrastus. In their attempts to restore the text to its original version, scholars struggled to match the Latin words in Pliny with their Greek and Latin counterparts in other scientific texts, as well as with the physical objects they denoted.[6] None of those involved in these efforts disputed the fact that errors existed in the *Historia naturalis*, but they disagreed as to their source. Were they the fault of earlier copyists, printers, and editors? Or could Pliny himself be blamed for these mistakes? At stake was the great reverence with which fifteenth-century readers regarded the Roman historian and the incontrovertible authority he wielded in matters of natural history.

Participants in this debate were few, but they numbered among a privileged class of physicians, lawyers, and scholars whose works were sponsored by the Italian aristocracy. The controversy began in 1492 when the Italian philologist Ermolao Barbaro (1454–1493) and Niccolò Leoniceno (1428–1524), a professor of medicine at the University of Ferrara, both published critical editions of Pliny's *Historia naturalis* unbeknownst to one another. Barbaro claimed in his *Castigationes Plinianae* (*Corrections to Pliny*) to have corrected over five thousand mistakes from the first several print editions of the work.[7] Leoniceno, too, identified errors in his *De Plinii et plurium aliorum medicorum in medicina erroribus* (*On the Errors in Medicine of Pliny and Many Other Medical Practitioners*); however, he attributed them to Pliny's mistranslations of Greek.[8] Finally, the following year, in 1493, the lawyer Pandolfo Collenuccio (1444–1504) published a defense of Pliny, countering Leoniceno's allegations by insisting that the Roman historian was a great botanist.[9]

Although Barbaro, Leoniceno, and Collenuccio disagreed on the nature of the textual inaccuracies in Pliny's *Historia naturalis*, they found common ground in the belief that these philological studies should be complemented with direct observation. Collenuccio maintained that in learning about nature it was not sufficient to "read authors, look at plant pictures, and peer into Greek vocabularies."[10] Similarly, Leoniceno admonished contemporary physicians who "unlike the ancients, do not go out into the field to study herbs for themselves" but rather rely on the observations of others.[11]

Comments such as these indicate that even before observation began to play a systematic role in nature studies during the 1530s and 1540s, scholars had started to draw on their experiences to evaluate classical knowledge. Pietro Bembo's *De Aetna* (*On Etna*) (1496), said to have been inspired by his close ties to Barbaro, showcases this new attitude. Conceived as a dialogue between Pietro and his father, Bernardo, the former describes his ascent of Mount Etna in Sicily. He informs his interlocutor that in the winter the whole mountain is covered in snow and that it remains frozen on the peak throughout the summer. To his father's rejoinder that the Greek philosopher Strabo recorded snow only in the winter, Pietro replies, "But firsthand inquiry (*experentia*) tells you that it lasts, as does practical experience, which is no lesser an authority (if Strabo will pardon me for saying so). And so, father, watch out again and again that you're not upset if I disagree with the ancient writers in any aspect of our present discussion of Etna. There's no reason that, at that time, all was not as they recounted it long ago, of which a great deal has persisted down to our own day, certain things have changed, and even some new factors have arisen."[12] At the same time as the young humanist—a correspondent of Oviedo's later in life—respects ancient learning, he accepts that the temporal distance separating the past from the present is not without its consequences, as the world has changed and new knowledge has been generated.[13]

The Spanish agronomist Gabriel Alonso de Herrera also advocates for the value of personal experience when confronted by written accounts at odds with direct observation. In *Obra de agricultura* (*Treatise on Agriculture*) (1513), which Oviedo cites in the *Historia general*, Herrera disagrees with Theophrastus on multiple points related to the planting and cultivation of various trees and plants on the basis of his hands-on learning.[14] Oviedo's critical readings of Pliny's *Historia naturalis* should be understood as emerging from the same intellectual currents as gave rise to these scholars' reevaluation of classical writings, which they increasingly began to understand as incomplete and in need of contemporary learning.

There is no indication that Oviedo was aware of the philological debates surrounding the *Historia naturalis*. His acquaintance with the Roman historian's work dates from roughly the same period, but given its large readership—a total of eighteen editions of the text had been issued by 1500 alone, fifteen in Latin and three in Italian—Oviedo could have easily encountered the tome in any number of private libraries in Italy and Spain, as it was a standard of many collections.[15] He did, however, hear the humanist Giovanni Pontano (1426–1503) lecture on another issue related to the textual imperfections in the work—namely, the question of its dedicatee. One of the first print editions of the *Historia naturalis*, edited by Giannandrea Bussi and published in Rome in 1470, addressed the work to Domitian, Emperor Vespasian's son. Three years later, an edition published by Konrad Sweynheim and Arnold Pannartz identified the recipient as Domitian's younger brother, Vespasianus or Titus. That no consensus had been reached on this question in subsequent decades is clear in Barbaro's inconclusive discussion of the topic in his *Castigationes Plininae* in 1492.[16]

In the *Historia general*, Oviedo takes a somewhat contradictory position on the issue: at the same time as he considers that Pliny could have written to either son, he stands by an earlier statement from the *Sumario* in which he identifies Domitian as the true dedicatee.[17] Oviedo justifies his stance, explaining that he had been present in Naples in 1500 to hear Pontano's reasoned arguments supporting this same position.[18] That Oviedo feels the need to defend his views on this seemingly inconsequential subject would suggest that he was cognizant of the polemical nature of the *Historia naturalis* in certain intellectual circles.[19]

Whatever the case may be, what leaves no room for doubt is the tremendous influence of the *Historia naturalis* on Oviedo's *Sumario* and *Historia general*. Pliny is also an indispensable reference work in Oviedo's genealogical and heraldic treatise, *Batallas y Quinquagenas* (1535–1557), in ways that will become more apparent in the next chapter. Indeed, so tight-knit and multidimensional is the relationship that Oviedo establishes between himself and his revered classical predecessor that an exhaustive analysis of it exceeds the scope of the current chapter. Rather, the connections between the two will be revisited and expanded on throughout this book to better elucidate their depth and complexity.[20] The following pages will explore how Oviedo complemented and corrected Pliny's *Historia naturalis* with information he acquired through direct observation in the Indies.[21] In so doing, Oviedo takes part in the new kind of hands-on learning for which Barbaro and Leoniceno had advocated at the turn of the sixteenth century, and which scholars such as Bembo and Herrera had already begun to put into practice. Unlike his contemporaries in Europe, however, Oviedo's subject was the relatively uncharted and unknown territories of the New World. Neither Pliny nor any other ancient authority had written of these lands, and most of Oviedo's readers would never encounter them personally. Oviedo faced the daunting challenge of establishing the credibility of a narrative centered primarily on his eyewitness testimony.

Although Oviedo was one of the first to employ classical knowledge as a measuring stick against which to evaluate his observations of the natural world of the Antilles, he would not have regarded his approach as original or groundbreaking. As he explains in the *Sumario*, he considers his research methods to be aligned with those of Pliny: "That which [Pliny] heard he attributes to who said it, that which he read he attributes to the authors who first made the observation, and that which he witnessed he gathered in the aforementioned *Historia*. I want to imitate him in all of these ways."[22] Like Pliny, Oviedo incorporated information into his writing from published authorities, the oral testimony of others, and his own personal experiences. However, whereas Pliny devoted the entirety of book 1 of the *Historia naturalis* to providing readers with an exhaustive list of authorities he had mined for facts, Oviedo was much more concerned with proving the veracity of his work. Unable to depend on a well-developed tradition of writing on the Indies and critical of the most comprehensive extant work on the subject, the Italian humanist Pietro Martire d'Anghiera's *De orbe novo decades* (*Decades of the New World*) (1516, 1530), Oviedo relied much more on his eyewitness observations than Pliny did. In his efforts to convince readers of the truth of his account, he emphasizes the struggles he experienced in acquiring this learning, contrasting the "two billion labors and hardships and dangers in the more than twenty-two years that I have seen and experienced in my own person" with the two million volumes that Pliny claimed to have read.[23]

To make himself better understood and endow his *Historia general* with the weight of classical knowledge, Oviedo establishes parallels between the flora and fauna he observes in the Indies and Pliny's remarks in the *Historia naturalis*. Oviedo relates what Pliny writes about the *physiter* in India because he surmises that it too is a whale, just like the sea creatures he has encountered in the New World.[24] In describing the tortoises on the island of Hispaniola, Oviedo acknowledges his habit of referring to Pliny in instances in which he feels readers might question his account: "Sometimes I look for witnesses in the ancient authors, so that readers believe me as a modern author speaking as an eyewitness to those who are far from these our Indies, because here all who are not blind see [these things]."[25] A case in point, Oviedo verifies his observations of these tortoises by setting them alongside Pliny's descriptions of those found in the Indian Ocean and the Red Sea. At the same time as Oviedo acknowledges that Pliny has much more information at his disposal than he does, he defends his account as being better substantiated: whereas the Roman historian appears not to have ever seen the tortoises about which he writes, Oviedo boasts that he has not just seen but even many times eaten those on Hispaniola.[26] As much as Oviedo respects Pliny as a historian and an investigator, he seeks to demonstrate to readers that his authority is not absolute.

The esteem with which Oviedo treats Pliny is not diminished by his admission that even Pliny needs to be emended occasionally. When Oviedo encounters disparities between what his own experiences lead him to believe and what Pliny reports in the *Historia naturalis*, he does not dismiss the latter's version; rather, Oviedo proposes many of his observations as updates to classical knowledge. In his discussion of how pearls are formed, Oviedo observes that the reasons given by Pliny and other classical authors to explain the gem's variations in color do not seem to apply to pearls in the Indies. Instead of discounting Pliny's assertions, however, Oviedo concludes: "It could be that in these parts [pearls] are formed and grow in one way and in the east and where Pliny and other authors say that there are [pearls] they are formed and grow differently. . . . Because nature in some parts operates differently in the same species of creature."[27] The shortcomings of the *Historia naturalis*, Oviedo understands, emanate from the ancient world Pliny knew and the ancient authorities he cited. They are not indicative of error but rather point to the fact that his work needs to be augmented.[28] In an attitude that recalls Bembo's words to his father in *De Aetna* (1496), Oviedo employs his own work to supplement Pliny's: "We should help [Pliny] write what he did not know nor found written about the southern and eastern parts of these our Indies as well as its other regions."[29] Oviedo completes Pliny's *Historia naturalis* in a "collaborative" project that brings together the ancient natural historian and his modern successor.

Oviedo's treatment of Pliny's *Historia naturalis* in the *Sumario* and the *Historia general* embodies changing early modern views toward the past. Just as the Plinian controversy of the 1490s exposed some of the problems inherent in the reception of classical learning in the Renaissance, so too the discovery of the New World demanded a reevaluation of ancient authorities. In writing a natural history on the basis of eyewitness testimony, Oviedo joins a community of scholars of the likes of Bembo and Herrera, who had begun to articulate the facts of a lived reality at odds with the wisdom of their Greek and Roman forefathers. Although these intellectuals continued to rely on Aristotle, Pliny, Dioscorides, and Theophrastus to give meaning to the natural world surrounding them, they became increasingly aware that, in and of themselves, these sources were insufficient. Oviedo's amicable, yet at times strained, relationship with his Roman predecessor reflects these developments.

The *Historia general* and Agricultural Traditions of the Early Renaissance

In *Batallas y Quinquagenas*, Oviedo's two literary alter egos, the elderly warden (Alcaide) and the youthful Sereno, discuss the military accomplishments of Spain's elite. Sereno hesitantly interrupts the older man's stories of heroism and daring to express his reservations at the human costs of such bloodshed. The Alcaide sympathizes with his companion's sentiments but argues that arms are vital to a prosperous society. He recalls a line from Xenophon's *Cyropaedia* (370 BCE) to this effect: "The valiant King Cyrus said that the art of warfare was no less necessary than agriculture."[30] By the time Oviedo penned these sentiments, sometime after 1535, the idea that agrarian life provided benefits comparable to those of commerce, law, and the military arts was commonplace.[31] Roman scholars and farmers of classical antiquity such as Marcus Terentius Varro (116 BCE–27 CE) and Lucius Junius Moderatus Columella (4–70 CE), both of whom Oviedo cites in the *Historia general*, spoke to the merits of country living. Other classical, medieval, and contemporary authorities with whose works Oviedo was acquainted—among them Marcus Porcius Cato (234–149 BCE), Virgil (70–19 BCE), Pliny, Piero de' Crescenzi, and Gabriel Alonso de Herrera—similarly reflected on the advantages of the farm and the fields. Oviedo's understanding of the natural world at large, and of New World flora and fauna in particular, is influenced in part by his readings of these authors and the value they place on the pleasures and utility derived from agrarian life.

The enormous editorial success that Pliny's *Historia naturalis* enjoyed in the last decades of the fifteenth century was matched by that of the Roman agricultural writers.[32] These authors strove to provide a learned perspective on farming practices, and the Renaissance inherited from them a time-honored respect for the task of managing a villa as a noble, gentlemanly occupation. Indeed, it was held as common knowledge that agriculture allowed mankind to improve on nature's offerings and transform them into bountiful sustenance. In *De natura deorum* (*On the Nature of Gods*), another work Oviedo references in the *Historia general*, the Roman philosopher Cicero describes the great power of human beings over their landscape: "We enjoy the mountains and the plains, the rivers and the lakes are ours, we sow the crops and trees, we give fertility to the land by conveying water to it, we confine the streams, we straighten or divert their course—in short, by means of our hands we endeavour to create in nature a kind of second nature."[33] Nature exists to serve human needs, but it falls to man to harness its possibilities.

Roman agricultural writers achieved mastery over this "second nature" by combining book learning with hands-on application; they did not just write about husbandry but practiced it on their own estates. Oviedo continues this tradition: *Book 7* of the *Historia general* reads as an agronomic manual of sorts, and the cultivation and harvest of island flora are also mentioned in books 8, 9, 10, and 11. In the case of corn (*Book 7*), much of Oviedo's knowledge of its planting and harvest comes directly from having cultivated the crop himself. At the same time, he cites Virgil, Herrera, and Pliny to lend credence to the agricultural techniques colonists have learned from the natives.[34] Much as Cato, Varro, and Columella wrote treatises informed by a combination of book learning, expert advice, and their own lived experiences, in the *Historia general* Oviedo advocates for the coupling of ancient learning with practical knowledge, and he cites these classical authors to that effect.[35]

In addition to his readings of these ancient Roman writers, Oviedo consulted several more contemporary farming handbooks. One of these, the Italian Piero de' Crescenzi's *Liber ruralium commodorum* (*Book of Rural Commodity*), initially came off the printing press in 1471, roughly at the same time as the first editions of Pliny's *Historia naturalis* and the Roman agricultural treatises.[36] The *Liber* quickly became the standard manual on the subject in Europe until the mid-sixteenth century, and it was reissued multiple times in France, Italy, and Germany.[37] It appeared on the bookshelves of several of the most prominent Italian intellectuals whose works Oviedo read or that he knew personally—namely, Petrarch and Leonardo da Vinci.[38] Given that the incunable edition of the *Liber* printed in Spain in 1476 did not meet with the same enthusiastic reception as elsewhere in Europe, it stands to reason that Oviedo encountered the work during his sojourn in Italy (1499–1502).[39] After all, Bologna, in northern Italy, was the *patria* of the medieval author and the locale where Crescenzi claims to have received his practical knowledge of farming.[40] It is possible that Oviedo had access to an illustrated edition of the text, such as the one published in 1495 in Venice. Although the vast majority of its woodcuts showcase the typical labors on a farm, several of them portray its avian and animal life.[41] These illustrations belong to a corpus of printed visual material depicting the natural world that set a precedent for the woodcuts Oviedo would include in the *Sumario* and the *Historia* (1535).

As the first comprehensive treatise on agriculture since the Roman agronomist Palladius's *Opus agriculturae* (*The Work of Farming*) in the fourth century, the *Liber* reflects its classical legacy at the same time as it anticipates the newfound importance that rural life would assume in the Renaissance, informing Oviedo's views on the subject. The vision of the natural world that predominates in Crescenzi's handbook is one in which utility takes precedence over enjoyment: over half of the *Liber* explores the benefits of trees and herbs to the human body. Even *Book 8*, devoted to pleasure gardens, calls attention to the fact that these spaces of rest and relaxation owe their existence to the produce reaped in the fields, vineyards, and orchards, the bulk of an estate's income.[42] In addition to their practical purposes, Crescenzi holds cultivated fields to be a source of beauty in and of themselves, which arises directly from their utility.[43] Oviedo's comments regarding the estate of Huerta del rey in *Batallas* are informed by a similar mindset; referring to the property as "a beautiful portion of cultivated land and profitable," Oviedo expresses admiration that such fields are both aesthetically pleasing and financially lucrative.[44]

The idea of the countryside as simultaneously a place of *negotia* and *otia*—of work and leisure—persisted into the Quattrocento. In an incomplete treatise on the villa written by Leon Battista Alberti (1404–1472), one of his interlocutors maintains that the primary function of a landed estate is self-sustenance balanced with the pursuit of "esthetic pleasures, good exercise, spiritual refreshment, and mental stimulation."[45] It follows that the ability to speak intimately and knowledgeably about the land and its bounty was the mark of a nobleman: in Cristoforo Landino's preface to his commentary on Dante, which Oviedo read, the Florentine humanist refers to his patron Lorenzo de' Medici admiringly as a "fine husbandman."[46]

Although this kind of praise harkens back to the ideals of ancient Rome, it acquired a greater urgency in the Renaissance in response to the emerging belief that somehow the benefits of rural living had gone unappreciated in recent times.[47] In his *Obra de agricultura* (1513), Herrera laments that Spaniards do not value agricultural work as much as the Romans and consequently are more ignorant in their knowledge of this art.[48] Oviedo, too, remarks in several instances on the overlooked advantages of agrarian life, complaining that Spaniards in the New World prefer to venture their luck in mining or trade rather than cultivate the land.[49] He contends that the body and soul enjoy great repose in farming, much more so than soldiers subject to the ambitions of their commanders.[50] In Italy humanists also advocated for renewed attention to the labors and fruits of the land: in Michelangelo Tanaglia's verse treatise *De agricoltura* (*On Agriculture*) (1483), he claims that the "cultivation of the soil rank[s] only behind letters and arms in the hierarchy of honor."[51] Similarly, patricians and intellectuals with whom Oviedo was acquainted celebrated in verse and prose man's interventions in nature and put into practice these values in their daily lives. It is to them that we will turn shortly.

The classical, medieval, and contemporary agronomists that Oviedo references in the *Historia general* are the literary foundation on which he constructs his vision of New World flora and fauna. In defending the labor of the land as a dignified enterprise, authors such as Cato, Varro, Virgil, Columella, Crescenzi, and Herrera established agronomy as a reputable subject of life and literature. Their emphasis on the need for man to exploit the soil defines Oviedo's approach to Spain's overseas colonies as well. Moreover, these agricultural writers doubled as actors, combining the knowledge they obtained via book learning with theirs' and others' lived experiences, as Oviedo would do. Despite the geographical and, in many cases, temporal distances between Oviedo and his predecessors, they serve as an important model for his attempts to document the natural world of the Indies.

Oviedo's Memories of Eden: The Early Renaissance Garden

In the *Historia naturalis*, Pliny credits the horticultural interests of his countrymen to the renowned gardens of antiquity—those of the Hesperides and of the rulers Adonis and Alcinous, as well as the hanging gardens—and he praises the former kings of Rome who "cultivated their gardens with their own hands."[52] Writing on the other side of the world thousands of years later, Oviedo too mentions gardens—not the legendary ones celebrated by his Roman predecessor, however, but rather several with which presumably he was directly acquainted, having served the patricians to whom they belonged. Speaking in praise of the pineapple, which was virtually unknown in Europe at the time, Oviedo declares its superiority over all the fruits in King Ferdinand I's Neapolitan gardens in the villa of Poggio reale, Ercole d'Este's island retreat of the Esquiva Noya, and Ludovico Sforza's transportable garden.[53] That the pineapple outrivals any of the fruits cultivated in these locales speaks not only to how much Oviedo values the New World fruit but also to key features of the gardens themselves. These horticultural spaces represented a status symbol for their patrons, an opportunity for them to showcase their wealth and power through the variety and abundance they displayed. Gardens appealed to Italian and Castilian elite for other reasons, too; following in the footsteps of the Roman kings of old whom Pliny lauds, they provided a place of work and recreation, of repose and study. Notwithstanding the importance of the founding of the botanical garden at the University of Padua in 1543 for the discipline of natural history, ultimately this event represents the culmination and synthesis of the kinds of horticultural activities that private citizens had long practiced in their residential gardens. Before Oviedo encountered the vast, untamed wilderness of the Americas, he had experienced the refined nature of the early Renaissance garden, and the courtly and humanistic attitudes and ideas that gave shape to it.

In many ways, gardens embodied the same return to the ideals of ancient Rome as other agricultural pursuits that have been explored in this chapter; indeed, the garden was an integral part of villa life, along with its fields, farm, orchards, and vineyards. Nevertheless, gardens were also conceived as independent entities—one did not have to own a country estate to have a garden, and plenty of them existed in urban settings. Be it in the country or the city, the garden was a private space, often enclosed by walls and located next to the owner's dwelling. Although it acted as a stage for certain leisure activities—dinners, meetings, and solitary strolls—one of the primary objectives of the garden was to provide sustenance for a family. Most contained fruits, vegetables, and herbs; in this sense, despite being set apart from the fields that surrounded them, gardens participated in the same delicate balance between *negotia* and *otia* characteristic of other tasks associated with rural life.[54]

During the sixteenth century, this emphasis started to change as ornate, lavish displays took the place of the simple vegetable plots and fruit trees of the early Renaissance.[55] Many of these villas were not built until the middle decades of the century; the Villa d'Este at Trivoli—famous for its statues, fountains, and greenery—was not completed until the 1560s. It is likely, then, that Oviedo did not experience this kind of decorative garden but rather its earlier manifestation as a place in which utility and pleasure coexisted, one a function of the other. In *Batallas*, Oviedo refers in passing to a variety of gardens and the activities that took place in them. He shares anecdotes regarding Pope Alexander VI's (1492–1503) strolls through the Belvedere garden; a banquet on the Marquis de Villena's estate; and the renown of don Jerónimo de Cavanilla's gardens in Valencia, whose size and abundance of fruits and water attracted foreign visitors.[56] It is clear from these remarks that Oviedo conceived of gardens as spaces of private reflection and social activity, as well as of beauty and productivity.

Oviedo's understanding of the garden and its functions was likely informed by the Moorish gardens of southern Spain. The most famed horticultural spaces on the peninsula, they had been in the possession of the Catholic Monarchs since the end of the Reconquest in 1492. The caliphs of Al-Andalus, as Spain under Muslim rule was generally known, viewed palace gardens as visual embodiments of their stewardship of the land; they sought exotic plants from India and China and created lavish displays of greenery and waterworks.[57] These horticultural interests were matched by the caliphs' scholarly patronage: the arrival of an illustrated manuscript of Dioscorides's *Materia medica* (*On Medical Material*), sent to Cordoba in 949 by Byzantine Emperor Constantine VII, heralded the emergence of the city as a center of botanical studies.[58] For the next three centuries, caliphs sponsored the writing of a variety of agricultural and gardening treatises based on some of the same classical sources that Oviedo cites, among them Columella.[59]

So greatly did the Catholic Monarchs value the gardens of Al-Andalus that they appear to have ordered no substantial modifications to them and even employed Morisco laborers and horticulturalists to cultivate them.[60] Scholars point to this aesthetic continuity as indicative of the fact that, at the time, Moorish gardens were not necessarily perceived differently than other Castilian gardens.[61] Indeed, by the end of the fifteenth century they had spread throughout Spain to varying degrees, albeit the dominant aesthetic in these horticultural spaces was classical. It is perhaps due to the ubiquity of such gardens that Oviedo does not single them out for comment in *Batallas* as did travelers to the region, such as the Italian Andrea Navagero (1483–1529), who praised those of the Generalife in Granada.[62] Instead, and as we have seen, Oviedo directs his attention to the renowned Italian gardens that so captivated his imagination.

The garden as described and experienced by Oviedo was a familiar notion to the Italian humanists with whom he was acquainted, who practiced horticulture alongside other intellectual activities. In engaging in these pursuits, they aspired to imitate their ultimate model, Petrarch (1304–1374). The Italian poet read Pliny the Elder, Pier de' Crescenzi, and Palladius, going so far as to annotate the latter's *Opus agriculturae* with two pages of handwritten notes detailing his experiments with the cultivation of various species in his gardens in Vauduse, Milan, and Arquà.[63]

Like their revered predecessor, many Italian scholars had private gardens in which they invested a great deal of care and attention, and foreign nature was also a topic of note in their travels. Nowhere is this interest more apparent than in the letters that Navagero exchanged with his son-in-law, the historian Giovanni Battista Ramusio (1485–1557), during the former's trip to Spain (1525–1526) as an ambassador of the Venetian Republic. The diplomat writes to Ramusio admiringly of the gardens, fields, forests, and rivers on his journey, populated with different plants and animals. He comments on some of the more unusual wonders to catch his eye, such as a kind of resin that in the springtime coats shrubs and bushes and stains sheep's fleece. Navagero also remarks on recently arrived phenomena from the New World—he encounters a dead, footless bird which he extols as one of the most beautiful things he has ever seen, and he tries a sweet potato (*batata*) and a fruit that he describes as having a flavor between melon and peach.[64] These diverse experiences constitute the better part of Navagero's written correspondence with Ramusio during his time in Spain, indicating the great enjoyment that he derived from these discoveries. Ramusio read the letters eagerly and shared them with their mutual acquaintance, Pietro Bembo, both hungry for such information.[65] Little wonder that Navagero returned to Italy with further news of New World nature in the form of Oviedo's recently printed *Sumario*, translated into Italian by the ambassador himself.[66]

Navagero's insatiable curiosity regarding these novelties did not diminish his nostalgia for his gardens on the island of Murano and at Selva. In his letters, he urges Ramusio to take good care of the plants, inquiring as to how much the laurel has grown and voicing excitement at the thought of returning home to find the trees well planted.[67] Anticipating his son-in-law's reaction to such concerns, he writes, "You will be amazed that with such occupations as I have that I concern myself with these trifles, as well they can be called; but don't be alarmed, because apart from my [ambassadorial] post, there is nothing that I think about more nor with greater pleasure. Therefore, help me, my dear Ramusio, with all of your care and diligence in this as the greatest and most important thing that you could do for me, and imagine that I am Epicurus and that I must spend the rest of my life in gardens."[68] For Navagero, the natural world is to be admired, explored, and cultivated. These pursuits are not independent of one another but interconnected: he sends Ramusio new plants from Spain, such as a *caronba* tree from Barcelona and the seeds of a sweet orange tree, to plant in his gardens and at the Villa Ramusio.[69] The presence of these exotic species ensures that his gardens reflect the diversity of the natural world both near and far, embodying the verses of the poet Serafino dell'Aquila that Oviedo quotes in the *Historia general*: "per tropo variar, natura è bella" ["nature is beautiful because it is so varied"].[70]

Although the correspondence between Navagero and Ramusio is contemporaneous with the publication of the *Sumario*, it reflects attitudes in keeping with those Oviedo encountered in Italy at the turn of the sixteenth century. One stop on Oviedo's itinerary, Naples, was particularly celebrated for its flora and fauna, thanks in part to the horticultural interests of the royal family. Oviedo, who served King Frederick of Naples in the months before his kingdom was wrested away from him on November 11, 1500, praises the hunting on the king's estate as being unrivaled by that of any other prince.[71] It is possible that the hunting grounds Oviedo so admires are the forests of the Poggio reale. Built in 1487 by the Duke of Calabria Alfonso II, the Poggio reale served as the summer residence of the Neapolitan royal family.[72] An anonymous poet who accompanied King Charles VIII of France on his invasion of Naples in 1495 commemorated its outdoor spaces in the poem "Vergiez d'honeur," in which he described its gardens—some ornamental, others dedicated to medicinal plants—as well as its orchard, farms, fishpond, and hunting reserve. Oviedo's reference to the Poggio reale in his description of the pineapple in the *Historia* (1535) comes over thirty years since he had last been in Naples, and eight years after the villa was destroyed in the aftermath of the sack of Rome in 1527.[73] Nevertheless, its gardens live on in his memories as a kind of Edenic paradise.[74]

In Naples, too, Oviedo made the acquaintance of a fellow nature enthusiast, Giovanni Gioviano Pontano.[75] Pontano's reflections on gardening express contemporary habits and attitudes that Oviedo would have seen on display during his travels. In the treatise *On Splendor*, Pontano instructs the "splendid" man to have gardens full of "exotic and rare plants, arranged with art and due care."[76] He advises such a person to "procure nurseries, as well as flocks of peacocks, pheasants, partridge and other birds . . . and all other things that are capable of giving the soul an honest pleasure."[77] Like the ancient agronomists, Pontano understands villa life as a tranquil retreat from the demands of politics and war, and he celebrates the human industry that transforms nature into this ordered refuge. Indeed, Pontano was a horticulturalist in his own right, this knowledge being expressed most completely in *De hortis Hesperidum sive de cultu citriorum libri duo* (*The Garden of the Hesperides, or The Cultivation of Orange Trees in Two Books*). Modeled after Virgil's *Georgics* (29 BCE), Pontano's poem pays homage to its ancient lineage at the same time as it presents new subject matter: the orange.[78] The Neapolitan poet recounts the mythical origins of this fruit in Italy and provides information on the cultivation of three of its varieties—bitter oranges (*Citrus aurantium*), citrons, and lemons.[79] The poem also includes the first known references to the sweet orange (*Citrus sinensis*), originally discovered on Vasco da Gama's voyage to India.[80] *De hortis Hesperidum* expresses pride in the autochthonous—according to the poem, the citrus is a gift that Venus bestowed specifically on Italy—while also embracing the foreign.

As a space that showcased exotic and rare plants, the early Renaissance garden became the ideal grounds for the experimental cultivation of New World species. In the *Historia general*, Oviedo mentions having seen corn growing on the estate of Hernán Ramírez Galindo in Madrid, adding that it was being cultivated as well in many parts of Andalusia.[81] On the other side of the ocean, he credits the priests in Santo Domingo with having successfully introduced plantains that they had acquired from the Canary Islands into their monastery gardens.[82] Before crops such as maize, tomatoes, potatoes, and tobacco were grown for mass consumption in fields or poked and prodded in university botanical gardens, these and other such foreign species were gifted and exchanged among wealthy patricians and humanists and planted in private garden plots.[83] Members of aristocratic families that Oviedo served, among them Isabella d'Este (1474–1539) and Ludovico Sforza (1452–1508), sent nonnative seeds, fruits, and vegetables to one another, and in their correspondence they provided instructions for their planting and food preparation.[84] The enthusiasm driving these individuals to acquire and cultivate novelty items anticipates the collecting impulse that emerged in scientific communities during the sixteenth century. In late fifteenth-century Spain and Italy, however, these practices were not yet institutionalized. Instead, they spoke to the interests of a cultured elite for whom gardens were as much a display of power as an aesthetic retreat and a place to exercise mind and body. During Oviedo's time in Italy, he experienced the lived reality of the garden—its botanical splendors, brought to fruition by the same hands as those that reimagined these Edenic spaces in verse and prose. These individuals contributed, in word and deed, to laying the groundwork for Oviedo's approach to the natural world in the *Sumario* and the *Historia general.*

Old Words and New Worlds

When Oviedo arrived on the shores of Castilla del Oro in 1514, he was greeted by exuberant, exotic, and untamed jungles. Although many of the flora and fauna of these regions were completely unlike anything Oviedo had ever seen, he was not unprepared for his encounter with them. Rather, his years immersed in the courtly culture of early modern Spain and Italy had given him the coordinates by which to navigate this new reality, which he ordered and described according to the intellectual currents then in vogue. With their balance between leisure and work, gardens put into practice the precepts of the agricultural treatises of Columella, Crescenzi, Herrera, and others. These, coupled with Pliny's *Historia naturalis*, provided the classical underpinnings for the approximations of Pontano, Ramusio, and Navagero to nature, both in its literary manifestations and as a physical reality. But, not content to rest on the laurels of their predecessors, these early humanists became increasingly attuned to the novelties surrounding them, which challenged them to find a place for their lived experiences among classical accounts that did not always easily accommodate their testimonies.

The *Sumario* and the *Historia general* reflect these developments. An ocean away from the comforts of home, Oviedo put into word and image a new universe of flora and fauna in ways that were both familiar and novel to his readers. By drawing on Pliny's *Historia naturalis*, Oviedo seeks to bolster his eyewitness claims with the weight of ancient authority. At the same time, he is ever conscious of the need to move beyond traditional wisdom to account for what he is able to perceive with his five senses, as are his contemporaries in Italy. Oviedo's experiences in early modern Europe at the turn of the sixteenth century impressed on him the beauty of nature as well as the need to capitalize on its potential as a source of sustenance and utility for humankind. These are the attitudes and perspectives that frame Oviedo's treatment of New World nature.

CHAPTER 2

Reality and Its Representation

Oviedo and the World of Renaissance Art

Only once in Oviedo's lifetime does he encounter a bird so beautiful it defies all descriptive possibilities of language or art. Given to him by Captain Andrés de Urdaneta upon the latter's return from the eastern Indonesian Spice Islands (the Moluccas) sometime after 1536, Oviedo praises the bird unreservedly: "Of all the birds I have seen in my life, this is the most elegant."[1] Yet he is at a loss to communicate even the most basic facts about the creature, including its name; Urdaneta has little familiarity with the species and Oviedo none at all. Studying the shrunken and dried specimen lying before him, Oviedo must conjecture on its size as well as its ability to fly. These difficulties, however, pale in comparison to the bird's utter majesty, which Oviedo insists no pen nor paintbrush could ever reproduce: "I confess that no painter could paint this bird, based on what I have said. . . . In my opinion, it is the most extraordinary in its feathers and beauty of all I have seen, and the one I have most admired. The bird is made with such craftsmanship and hand that it is possible, and should be believed, that such an artist spared no talent on this bird. Neither could any painter, sculptor, or orator express so naturally, nor could human ingenuity make perfectly understood what these birds are like."[2] Oviedo's encounter with this beautiful bird exposes a dilemma at the heart of his attempts to render visible New World nature: the insurmountable division between reality and its representation. At the same time as he admires the supreme talent that created the bird—putting on display the Renaissance topos of God as the ultimate architect and artist—so too does he despair at the inadequacies of human expression to describe it. In comparing divine and human artistry, Oviedo reveals the basic criteria he employs to judge works of art: the ability to accurately reproduce external reality. Although in this instance Oviedo deems such attempts insufficient, nevertheless he echoes the sentiment of his age in establishing mimesis as the measure of artistic genius.

This episode with the bird is not the first nor the last time Oviedo invokes the power of the visual arts in his writings on the Indies. Indeed, more than any other early New World chronicler, Oviedo's evolving approach to depicting nature increasingly embraced a pictorial component. The *Sumario* (1526) contained four woodcuts of flora and fauna; nine years later, the first part of the *Historia* (1535) included twenty-six woodcuts. By the time Oviedo deposited the three-part manuscript of the *Historia general* in a Sevillian monastery for safekeeping until such a time as it could be published, in 1549, the work had eighty-two hand-drawn images.[3] These illustrations comprise one of the most studied aspects of Oviedo's writings on the Indies, and for good reason: they are the first to capture New World nature based directly on eyewitness observation and testimony.[4] Over time, these empirical methods would become a defining feature of Spain's efforts to learn about its overseas possessions.[5]

Although the woodcuts and drawings in the *Sumario* and the *Historia general* might be the clearest indication of Oviedo's sustained interest in graphic representation, they are certainly not the only ones.[6] Rather, Oviedo's initial enthusiasm for this topic emerged from his experiences as a young man in the courts of early modern Spain and Italy. It was here, and primarily in the courts of Milan, Mantua, and Naples, that Oviedo acquired the fundamentals of his education in the world of Renaissance art. The aesthetic criteria that he developed as a result of his interactions with the foremost art and artists of the time, as well as his readings of Pliny and Petrarch, constitute the foundation of his understanding of this topic.

Oviedo references his encounters with the artistic milieu of Renaissance Italy several times in the *Historia general*, but it is in *Batallas y Quinquagenas* that he most elaborates on these experiences. Conceived as a conversation between Oviedo's two alter egos, the youthful Sereno and the elderly warden (Alcaide), the work surveys the genealogies and heraldic devices of Spain's elite, many of whom Oviedo was acquainted with. He did not begin composing *Batallas* until 1535, and it remained unfinished upon his death in 1557. Scattered throughout the work are Oviedo's reminiscences of his time at the court of the Catholic Monarchs (1490–1497), in Italy (1499–1502), and in Spain after his return and during his intermittent visits there from the New World (1503–1513, 1515–1520, 1523–1526, 1531–1532, 1534–1535, and 1546–1549).

Via the figure of the Alcaide, Oviedo recounts anecdotes and pithy statements about the visual arts, recalls conversations that he sustained on the subject, and describes works of art—mainly paintings and portrait medals—that he saw.

This chapter contributes to ongoing scholarly efforts to contextualize the woodcuts and drawings in Oviedo's writings on the Indies within the sphere of sixteenth-century visual culture. In *Naturaleza e imperio* (2003), Carrillo Castillo elaborates on Oviedo's knowledge of heraldic conventions; the changes rendered to the woodcuts in the *Sumario* at the hands of its most enthusiastic contemporary Italian reader, Giovanni Battista Ramusio (1485–1557); and the relationship between Oviedo's verbal descriptions of medals and paintings and the rhetorical and poetic tradition of ekphrasis. In contrast, I explore Oviedo's familiarity with the conventions of Italian Renaissance art as evidenced in his multifaceted engagement with the subject in *Batallas*. These experiences, and the knowledge that Oviedo gained as a result, would provide the framework for his later attempts to illustrate New World nature.

Art at the Court of the Catholic Monarchs (1490–1497)

The reign of Queen Isabella and King Ferdinand (1479–1504) in Castile was one of economic prosperity and royal patronage of the arts.[7] The queen's rich, embroidered fabrics and ornate jewelry caused much admiration among her subjects and visitors from abroad.[8] These fashions were emulated by members of the court and nobility, who procured other luxury items as well: tapestries, books, sculptures, paintings, and engravings.[9] Artists and artisans from northern Europe began to migrate to Spain in ever-increasing numbers during the second half of the fifteenth century, heralded by the earlier visit (1427–1428) of the Flemish artist Jan van Eyck to Spain and Lisbon as an emissary for Philip the Good, Duke of Burgundy.[10] These new arrivals received commissions primarily for items of religious devotion, such as altarpieces, chapels, and funerary monuments.[11] Castile was the center for much of this activity and the Catholic Monarchs its principal patrons.[12] By the time Oviedo arrived at court in 1490 as a *mozo de cámara* for Prince John, Isabella had in her employ several court painters, and she would hire at least one more during Oviedo's time there.[13] Yet, despite these developments, Oviedo makes few allusions to the artistic culture of the Castilian royal court in his writings.

Although he comments admiringly on the attire and accessories of the upper echelons of Spanish society in *Batallas*, he does not name those responsible for their creation, nor does he single out particular encounters he sustained with Hispano-Flemish artists or their art. Conversely, the artists Oviedo met in his travels around Italy are subjects of note in the *Historia general* and *Batallas*. This disparity is indicative of the different attitudes that existed in early modern Spain and Italy toward art and its makers. We begin our discussion in Spain, to better understand these differences and to set the stage for Oviedo's first encounters with Renaissance Italy.

Oviedo's arrival at the court of the Catholic Monarchs around 1490 coincided with the apogee in Castile of Hispano-Flemish art, a syncretic mix of Franco-Flemish and local aesthetic influences.[14] Isabella's preference for northern European art reflected the kingdom's long-standing commercial and political ties to this region, which were further solidified during her reign through the marriages of Prince John to Margaret of Austria in 1496 and Princess Juana to Philip the Fair the following year.[15] The predominant aesthetic throughout Spain in the fifteenth century, Oviedo would have been accustomed to seeing the hallmarks of the Hispano-Flemish style in any number of churches, its emotional intensity coupled with a lack of concern for technical refinement.[16] The popularity of this kind of art minimized the impact of the Italian Renaissance in Spain, a circumstance that would not begin to change until roughly the close of the fifteenth century.[17] Although the queen owned a handful of Italian paintings, the vast majority of her collection consisted of works by northern European painters and local artists schooled in Hispano-Flemish techniques.[18] The same holds true of the painters she engaged at the royal court: the first such appointment, made in 1480, was awarded to Francisco Chacón, a Spanish-born Hispano-Flemish artist.[19] The portraitists Antonio Inglés and Michel Sittow (1469–1525), both present at court in the 1490s, originated from northern Europe, as did Juan de Flandes, documented in the employ of Isabella for the first time on October 27, 1496.[20]

Little documentary evidence exists as to the responsibilities and commissions of these royal painters, but presumably they resided at least occasionally at court. There is no evidence that Oviedo was personally acquainted with them, however; the size of the royal court had swelled to over a thousand people by 1492, and such encounters were not inevitable.[21] It stands to reason that Oviedo was better acquainted with Prince John's attendants, having been one himself. The prince's entourage included an official painter,

Jaime Serrat from Saragossa, whom Isabella had appointed in 1487 as a "painter, silversmith, and lance maker."[22] Yet in the *Libro de la cámara real* (*Book of the Royal Chamber*), composed in 1546 for Charles V's son Prince Philip, Oviedo makes no reference to any responsibilities associated with a court painter. Instead, he describes the embroiderer and the silversmith, lauding them as "great professions, and they were busy most of the time because the prince my *señor* was much inclined to dress richly, and they were favored and well paid and lodged."[23] Nor does Oviedo mention Serrat by name, despite identifying a great many others in the prince's employ. The municipal council of Saragossa appointed Serrat painter of the city and the archbishopric on September 29, 1492, and it is not known if he continued to serve the prince during the intervening years, when Oviedo was at court.[24] Complicating matters further is the fact that Oviedo did not write the *Libro* until nearly fifty years after the prince's premature death on October 4, 1497; the absence of both the position of official painter and the person of Serrat from the work could be due to lapses of memory.

Nevertheless, I suggest that the veil of silence in Oviedo's writing regarding the presence of artists at the royal court points to the realities of life there, as well as to deeply rooted beliefs concerning the role of the artist in Castilian society. A defining feature of the monarchy was its itinerant court, which prevented the king and queen from establishing a "fixed royal image through collection or household furnishings" or becoming a center for artistic and cultural activity akin to the Italian courts of Milan, Mantua, and Naples.[25] Moreover, while Isabella was a patron to artists, sculptors, architects, and illuminators, the extent of her involvement in these commissions is unclear; for most of her reign, her attention and monies were directed toward military matters.[26] Inventories of Isabella's possessions taken after her death in 1504 put on display the "random accumulation" of medieval collecting rather than the deliberate and purposeful engagement with artifacts that marked the Italian Renaissance.[27]

The eclectic nature of Isabella's collections is suggestive of the status of the arts in Spain at the turn of the sixteenth century, where painting was still considered a manual labor and the artist, a craftsman.[28] In contrast, in Italy, Leon Battista Alberti's treatise "On Painting" (1435) had elevated painting from the artisanal activity it had been in the Middle Ages to an intellectual one, subject to the laws of geometry and the classical categories of rhetoric.[29] By the time Oviedo arrived in Italy, in 1499, Leonardo da Vinci too had begun to advocate for painting as a liberal art, alongside grammar, logic, rhetoric, geometry, arithmetic, astronomy, and music.

The great regard that Alberti, Leonardo, and other Italian intellectuals fostered for the visual arts resulted in a newfound respect for the figure of the artist. The socioeconomic possibilities of those who practiced this profession in Italy so impressed the Spanish painter and sculptor Alonso Berruguete (ca. 1488–1561) during his residence there (ca. 1506–1518) that upon his return to Spain he sought similar treatment. In so doing, he became the "first Spanish artist to promote himself without modesty."[30] Although Berruguete's efforts met with success—the soon-to-be emperor Charles V awarded him the title of Painter to the King (*pintor del rey*) in 1518—the idea of painting as an intellectual activity was far from universally accepted at the time, let alone twenty years earlier in the Isabelline court. Indeed, so mixed were opinions on this topic that it continued to be debated throughout the Spanish Golden Age (1500–1700).[31] In short, the visual arts were still widely regarded as a craft in Spain in the last decade of the fifteenth century, not the revered profession they were shaping up to be in Italy.

Like other royalty in Western Europe, the Catholic Monarchs prized sculpture, architecture, and painting for political and religious reasons. With 80 percent of the population illiterate, these artistic mediums were indispensable for communicating quickly and effectively with their subjects.[32] The heraldic devices of the royal couple—appearing prominently on clothing, buildings, and coinage, and featuring as well on the tombs, convents, churches, and architectural monuments that Isabella commissioned—effectively became their language of rule. Yet the monarchs' attitude toward these works of art and the artists who executed them was still largely medieval, incarnating values at odds with the burgeoning respect in Italy for the painter and his craft. So too, the prevailing Hispano-Flemish aesthetic, evidence of the close-knit relationship of Castile with its northern neighbors, slowed the arrival of Italian humanism to the kingdom. Little did Oviedo know when he embarked for Italy in 1499 that this would soon change: in the subsequent decade, a handful of Spanish artists would make the trip to Italy as well, eager to receive training from the masters there. Once back in Spain, they began to introduce this technical knowledge into Hispano-Flemish art. Like these artists, it was in Italy where Oviedo would be exposed to Renaissance art for the first time. The profound impact these experiences had on him—to the extent that he recalls them thirty-five years later in *Batallas*—is an indication of the vastly different world he left behind in Spain.

Oviedo's Italian Renaissance (1499–1502)

Oviedo arrived to an Italy in the heyday of its Renaissance. The artists Oviedo met—among them Leonardo da Vinci (1452–1519) and Andrea Mantegna (1431–1506)—were well aware of their participation in a pictorial tradition with a long and august lineage. In the prologue to his commentary on Dante, from which Oviedo cites in the *Historia general*, the Italian humanist Cristoforo Landino (1424–1498) traces this genealogy for his readers, beginning with its more immediate predecessors in Giotto (1267–1337) and Masaccio (1401–1428) and stretching all the way back to the arts of antiquity.[33] The language Leonardo, Mantegna, and their critics used to describe the artist's task had roots equally as deep: Pliny the Elder's (23–79 CE) descriptions of Greek and Roman painting and sculpture in the *Historia naturalis* provided an account of artistic excellence that was later endorsed by Petrarch and Boccaccio in *De remediis utriusque fortunae* (*Remedies for Fortune Fair and Foul*, 1354–1366) and *Genealogia deorum gentilium libri* (*On the Genealogy of the Gods of the Gentiles*, 1360), respectively.

The value these early Italian humanists ascribed to artists and their art set the standards for their treatment in the Quattrocento. Art in the courtly circles that Oviedo frequented was not just something aesthetically pleasing to look at but also a subject to be discussed, judged, and written about. The measures of "good" art—above all, the striving for verisimilitude, to the point of confusing the real with its reproduction—were well known to artist and beholder alike, and this common vocabulary became the basis of a lively exchange of ideas and anecdotes at court and on paper. *Batallas* too, participates in this collective enterprise: Oviedo's references to his readings of Pliny and Petrarch, the exchanges he sustained with Italian artists, and the artistic norms of Renaissance art attest to his involvement in this tradition.

Oviedo's Readings of Pliny's *Historia naturalis*

Pliny is a much-cited figure in *Batallas*, just as he is in the *Sumario* and the *Historia general*, though for much different reasons. Representing the only extant account of Greek painting and sculpture, of which no examples survived, the *Historia naturalis* served as a manual for early modern humanists and artists alike.[34] Pliny's numerous anecdotes involving these long-lost works of art and their creators informed contemporary readers of "what materials and techniques to use, what subjects to portray, what styles to emulate. It showed them how to comport themselves as artists, collectors, and viewers, and explained how to judge works of art."[35] Just as books 33 to 36 of the *Historia naturalis* were a necessary point of departure for humanists' inquiries into classical art, so too for Oviedo. He references these sections of Pliny to inform his biographical sketches of Spanish nobility in *Batallas*. In applying Pliny's anecdotes regarding the artistic culture of ancient Greece and Rome to contemporary subjects, Oviedo engages in the literary tradition initiated centuries earlier by Boccaccio of adapting ancient anecdotes to early modern figures and events.[36] Oviedo modernizes the ideas of his Roman predecessor, demonstrating their continued relevance and applicability while simultaneously championing Renaissance values and art.

Pliny structures his account of the history of art as a series of technical innovations that paved the way for the representation of an increasingly lifelike subject. These efforts began with "tracing lines round the human shadow" and gradually acquired greater complexity with the use of color, light, and shading.[37] In *Batallas*, Oviedo uses his understanding of these origins of art to inform an encounter he describes between don Alonso de Silva and one of his pages. The Alcaide tells Sereno,

> You ought to know that don Alonso often said to those who served him that a page who did not have any special skill was worthless; and one of the most principal of these skills was that of drawing well. So it happened that, one night, when don Alonso was writing by the light of a candle, his shadow was cast onto a white wall near where the page was waiting for his lord to finish the task at hand, to snuff the candles and do whatever else was requested. Because it was winter there was a fire in the hearth. As the page longed to learn how to draw, he took a piece of coal, and over on the white wall where the shadow of his lord's face and long nose was projected, he began to outline its profile, drawing the mouth identical to that of the shadow. Being absorbed in this task, and having almost completed the big head and monstrous picture of his drawing, the lord turned toward the page and saw what he was doing. And the page told him, "Don't move yet, my lord, I'm almost finished." His lordship, laughing, said, "Why didn't you tell me before what you were doing, you silly boy? If I were to measure my nose with a mirror, it wouldn't be as big as it is in this drawing. Finish it, finish it, for this was the origin of painting in the world." And

> it was true, the lord sent the page afterwards to Flanders and paid for him to learn how to paint, and he emerged from that experience an elegant painter.[38]

Despite the page's utter failure to capture his master's likeness by tracing his shadow on the wall, don Alonso recognizes the boy's potential in his instinctual use of techniques dating from the very beginnings of painting.

Readers of the period might have interpreted this episode as a tongue-in-cheek allusion to an anecdote from Pliny about the origins of modeling portraits from clay. The Roman historian explains that the Greek potter Butades invented a method of making relief portraits after his daughter, in love with a man about to go abroad, "traced the profile of his face, as thrown upon the wall by the light of the lamp."[39] Witnessing this, Butades pressed clay into the outline and then cast the relief into the fire. While Pliny's account focuses more on the novelty of the artistic techniques Butades employed rather than the resulting final product, one could imagine that the cast relief reflected distortions like those in the drawing of don Alonso's page. In any case, Pliny presents the potter's innovations as a preliminary step to a more lifelike portrait: his successor, Lysistratus of Sicyon, created casts by fitting plaster directly onto the model's face to give "faithful likeness" to the human figure.[40] While leaving much to be desired, the efforts of don Alonso's page and the Greek potter are praised by Oviedo and Pliny, respectively, as the foundation of future artistic improvements. In recounting this humorous episode, Oviedo alludes to one of the main criteria he employs in other instances in *Batallas* and the *Historia general* when commenting on works of art—namely, the importance of artistic naturalism, including accurate proportions, a discerning use of color, and direct observation of natural phenomena.

Pliny celebrated mimesis as the epitome of artistic genius as well, and he praised those artists who were able to "deceive" man and beast by faithfully rendering their subjects. He recounts an anecdote about the painter Apelles of Kos in which horses neighed at a horse he had painted as if it were real. For this feat, and others similar to it, Pliny refers to Apelles as a painter who "surpassed all the other painters who either preceded or succeeded him."[41] Another such account describes a painting contest between Parrhasius and Zeuxis: the latter "represented some grapes, painted so naturally that the birds flew towards the spot where the picture was exhibited," whereas the former "exhibited a curtain, drawn with such singular truthfulness, that Zeuxis . . . haughtily demanded that the curtain should be drawn aside to let the picture be seen."[42] Zeuxis reluctantly recognized his competitor's superior skill, as the painting had eluded even his discerning eye.

In *Batallas*, Sereno endorses Pliny's praises of artistic realism, saying, "I think that the better painting is one which most resembles the thing itself."[43] The Alcaide agrees, offering an example he had seen in Barcelona in 1519: a portrait of the Countess Ana de Cabrera, the wife of don Fadrique Enríquez II, the Admiral of Castile, painted by the Spanish artist and sculptor Alonso Berruguete. Describing the portrait as one of the finest in Berruguete's oeuvre, the Alcaide recounts that he was startled to see how perfectly it resembled her person.[44] He expresses even more astonishment at Berruguete's sculpted creations, maintaining that they are no less excellent than "the sculpture of Olympian Jupiter that Phidias made, nor Praxiteles's sculptures, nor all the work that Scopas, Bryaxis, and Leochares did for the memorable sepulchre of King Mausolus, nor what Apelles and Protogenes painted."[45] When Sereno conveys his surprise at hearing his interlocutor equate modern and classical art, the elderly man retorts: "Similarly, I am struck that, wise as you are, you think that I should hold in greater esteem a painting I have never seen but only read about in accounts singing its praises than one which, gazing upon it, seems to represent its subject matter so well. [And this is as it should be,] seeing that, ordinarily through experience, all the arts grow and perfect themselves by use and exercise."[46] The Alcaide calls attention to Sereno's faulty logic in privileging written descriptions of classical art over contemporary works, whose merits can be judged by the spectator. His scolding of the young man calls attention to how Oviedo employs Pliny's arguments to move beyond the prestige of classical art and advocate for Renaissance artistic production. By echoing the Roman historian's sentiment regarding art's gradual evolution toward refinement, Oviedo argues that modern achievements should be favorably compared to those of the ancients.[47]

As much as Pliny, and later Oviedo, touts imitation as the primary objective of artistic production, both acknowledge the impossibility of fully capturing external reality. Such is the case of the Greek artist Timanthes's painting of the moments preceding Agamemnon's sacrificial offering of his daughter, Iphigenia, to the goddess Artemis. Pliny describes the mournful atmosphere of the painting: "Upon the countenance of all present, that of her uncle in particular, grief was depicted; but having already exhausted all the characteristic features of sorrow, the artist adopted the device of veiling the features of the victim's father, finding himself unable adequately to give expression to his feelings. . . . Indeed, Timanthes is the only one among the artists in whose works there is always something more implied by pencil than is expressed, and whose execution, though of the very highest quality, is always surpassed by the inventiveness of his genius."[48] Although Pliny commends Timanthes's depiction of the grief of Iphigenia's uncle, he most admires Agamemnon's hidden profile. The Roman historian applauds the artist for having produced a painting whose significance transcends its physical representation on the canvas to allude to an emotional profundity implied but not expressed. In so judging the results of Timanthes's composition, Pliny points to the need to recognize the limitations of pictorial reality, as even the most skilled artist cannot portray all the nuances of human emotion.

Oviedo welcomes similar praise in his writing, portraying himself as a Timanthes-like figure in recognizing his limitations as an artist and writer. He admits openly when a given task exceeds his artistic or rhetorical capabilities. A case in point is his description of the welding tree in the *Historia general*; so extraordinary is its appearance that he finds himself at a loss to depict it verbally or visually. Instead, he defers to such artists as Alonso Berruguete, Leonardo da Vinci, and Andrea Mantegna to paint the plant.[49] Similarly, he invokes Pliny's account of Timanthes's painting explicitly in *Batallas* to explain the difficulty of putting into words the suffering of the Duke of Calabria after his father was forced to abandon the throne of the kingdom of Naples in 1501. The Alcaide complains to Sereno:

> It seems the more time it takes to finish these dialogues, the more occasions abound to detain ourselves, so much so that I would fear ever reaching the conclusion if not for the remedy that some ancient painters used. Namely, when their arts were not enough, they used the trick of painting the backs of images, wanting to give the impression that these were more beautiful than those images clearly displayed. And it is also fitting what Pliny says about that great painter Timanthes who, painting many [people] in sorrow, covered the face of the saddest among them, the father of Iphigenia, who was to be sacrificed. For this reason, Pliny says in praise of this painter that more was seen than what he had painted in all of his compositions. In this way, you [Sereno] have to understand, regarding what I say of the misfortunes of His Serene Excellency don Fernando de Aragón, Duke of Calabria, that his loss is much greater than what I know how to paint or say, without comparison.[50]

Later, when transcribing the amorous verses the duke had dedicated to his second wife, the marchioness of Zenete and Ayora, the Alcaide echoes these earlier sentiments, telling Sereno, "The duke's verses are more profound than they sound, imitating the works of that painter Timanthes."[51] In both instances, Oviedo employs the conventional analogy between painting and writing to equate these penned sentiments with Timanthes's decision to veil Agamemnon's face.[52] Be it painting, poetry, or prose, any form of human expression offers only a partial window onto its subject.

Petrarch's *De remediis utriusque fortunae* (1354–1366) and Humanist Art Criticism

While Pliny's *Historia naturalis* provided a repository of ancient artistic practices and skills for Renaissance Italy, it was Petrarch (1304–1374) who drew renewed attention to them by reinterpreting their significance in light of humanistic sensibilities and Neoplatonist concerns.[53] Petrarch engages with these topics most explicitly in chapters 40 and 41 of *De remediis utriusque fortunae* (*Remedies for Fortune Fair and Foul*), which constitute the most substantial discussion of art in the Trecento. In the work, the Italian scholar gives an account of his views on painting and sculpture through a conversation between Gadium (Joy) and Ratio (Reason). Although the differences of opinion between the two interlocutors make it difficult to interpret their exchange, in general Joy's laconic, insubstantial remarks on the pleasure to be had from the visual arts do not hold up to Reason's lengthy rejoinders.[54] At the same time as Reason does not entirely reject the merits of painting and sculpture, he cautions against the dangers of viewing works of art without proper understanding or knowledge.

The guidelines Reason proffers to counter the perils of Joy's unrestrained enjoyment of the arts are informed by the central tenets of Neoplatonism, with its distinction between the material and spiritual realms. These aesthetic principles inform Oviedo's understanding of the function and shortcomings of the visual arts as well. Oviedo does not engage directly with Petrarch's statements concerning art and the beholder from *De remediis* in his writings, but a reference to the work in the *Historia general* indicates he was acquainted with it.[55] At the very least, Oviedo shares with his humanistic predecessor a conception of art as an imperfect reflection of the divine, indicating the effect of *De remediis*'s philosophical outlook on the humanist art criticism of Oviedo's era.

During the conversation between Joy and Reason in *De remediis*, the former repeatedly declares his delight at painting. Reason rebukes his companion, arguing that the only kind of pleasure one should receive upon viewing a work of art does not have as its object any kind of material display but rather ethereal truths. He accuses Joy of taking

> delight in the pencil strokes and colors which please because of price and skillfulness—their variety and artistic composition. And you are fascinated by the lifelike gestures, the movement in these inanimate and immobile pictures, the faces jutting out of posts, and the portraits that seem about to breathe and make you think that they might utter words. The danger here lies in the fact that great minds, in particular, are captivated by these things—and what a peasant will pass off with brief enjoyment, a man of intellect may continue to venerate with signs of admiration. . . . Thus, when these fictions and contours in feckless colors delight you too much, turn your eyes to Him, who painted feelings on the face of man, intellect on man's mind, the stars on the heavens, the flowers on the earth—and you will disdain the paintings you admire.[56]

Reason contends that an aesthetic appreciation of painting must be undertaken with care and mindfulness. Mimesis is a double-edged sword: the epitome of artistry, this ability to reproduce external reality has the power both to serve as a vehicle of religious sentiment and to mislead the spectator into venerating false idols, celebrating the visual arts for what they are rather than what they represent. Through Reason, Petrarch advises readers to look beyond a painting's physical attributes and economic value to contemplate a higher spiritual reality.[57] He maintains that as imperfect representations of God's perfect creations, paintings are valuable only as tools that enable the viewer to gain a greater appreciation of the divine.

The Neoplatonic division between the material and the spiritual that permeates Petrarch's discourse on painting and sculpture in *De remediis* is integral to Oviedo's approach to his task as a historian of the Indies. The idea of God as the ultimate artist is nowhere more clearly articulated than in the episode with the beautiful bird that opened this chapter. Whereas Petrarch warns against worship of the painted or sculpted image, Oviedo has no such scruples about the godly artistry that is his primary object of study. Be it this unknown, utterly indescribable bird from the Spice Islands or any other phenomenon, Oviedo casts his eyes eagerly on the natural world surrounding him, convinced that there is no more laudable occupation as "seeing with one's corporal eyes all that is visible to them in this composition."[58] Faced with the majesty of God's creation, he acknowledges the failure of any attempt to portray these wonders—they are better seen firsthand than written about or illustrated.

Occasionally, Oviedo treats these inadequacies of human expression in a lighthearted manner. In one such instance in *Batallas*, the Alcaide recounts to Sereno an anecdote regarding don Hugo de Cordona. He recalls,

> Don Hugo was much in love with a very beautiful and noble lady, who I saw many times in the city of Naples. Although he had procured many excellent painters to paint her portrait, none of them was able to produce one as perfect as he desired. In spite of this, however, for his devotion and as a diversion he stored these portraits reverently in his chambers, some of them [in which her expression was] happy, others enraged, some in which she was well-dressed with her hair swept up, others with her hair down and in different styles. And one day, when I was laughing with him about these portraits, saying, "This version appears more like her than that one, and this one more than that," I asked him what he had decided to do with all of these "unresponsive women." With a deep sigh, don Hugo responded, "Only God knows how much more I would love to have the living woman than the one depicted in these portraits, but it would be a crime to harm them or fail to appreciate them, because they were made to imitate she who is within my soul." And saying this, he put his hand on his chest.[59]

Oviedo presents don Hugo's attempts to reproduce an exact likeness of his beloved with a tongue-in-cheek humor. Despite the obvious futility of the endeavor, don Hugo continues to commission portraits that fail to capture the noble lady's true essence. The myriad paintings crowding his chambers convert them simultaneously into a place of domestic tranquility and lovelorn torment. For the Alcaide's interlocutor, Sereno, this anecdote speaks to the superiority of the natural over the artificial, an opinion he summarizes as follows: "That which is natural is better than artificiality, and the difference between them is like that existing between death and life, between a truth and a lie, and between shadow and certainty."[60] In articulating these hierarchies, Oviedo establishes boundaries delimiting the influence of the pictorial and plastic arts. In the spirit of Petrarch, Oviedo affirms that a representation of reality is never equivalent to reality itself.

Pliny's *Historia naturalis* and Petrarch's *De remediis* set the standards for humanist critiques of art in the Renaissance. The most fundamental of these criteria was Pliny's vision of the history of art as constantly evolving over time, a progression that he ascribed to attempts to achieve verisimilitude, the essence of artistic genius. The anecdotes Pliny recounts about Greek artists who sought to confuse what they produced on canvas and sculpted in stone with physical reality provided the impetus for Renaissance artists to continue to aspire to these heights. Unlike these ancient artists, however, who looked to best nature and each other, Renaissance painters and sculptors competed not only against these standards but also against the accomplishments of their renowned predecessors. As Petrarch had done in *De remediis*, early modern humanists strove to articulate the artistic values of this august past in a Christian framework. Oviedo applied these principles to episodes in *Batallas* involving colorful contemporary figures—don Alonso, the Duke of Calabria, and don Hugo. Yet Oviedo's forays into the world of Renaissance art were not strictly literary in nature. Rather, they had their counterpart in the artists he met in Italy and the art he observed there. We consider these now.

Oviedo and the Artists and Art of Renaissance Italy

Of the four years Oviedo spent in Italy, it was his time at the courts of Milan, Mantua, and Naples that most defined his experiences there. Famed as centers of cultural and intellectual life, their rulers surrounded themselves with notable poets, artists, musicians, and historians. Oviedo's sojourn at these residences brought him into contact with some of the most innovative thinkers of his era, among them Leonardo da Vinci and Andrea Mantegna. Regardless of the material with which these artists worked—metal, gems, stone, or paint—they shared a common understanding of their task as grounded in the verisimilitude for which Pliny had advocated. In contrast to natural philosophers and medical doctors who were only gradually beginning to employ direct observations in their studies by the end of the fifteenth century, these artists had long ago turned to the natural world as a model for their creations. Those who engaged with external reality most deeply were the most famed of their day; these also happen to be the artists with whom Oviedo became acquainted in Italy and whose exploits he praises in *Batallas*. These experiences allowed Oviedo the opportunity to view the visual embodiment of the artistic values held so dear by contemporaries and classical predecessors alike. Once in the Indies, Oviedo would put into practice many of the strategies employed by these artists, showing the extent of his debt to them in his attempts to render visible New World nature.

Leonardo and Caradosso: Art at the Court of the Duke of Milan (1499)

Oviedo's first stop in Italy was Genoa, where he served one of the governors, Giovanni Adorno. Giovanni ruled the republic with his brother, Agostino Adorno, on behalf of the Duke of Milan, Ludovico Sforza (1452–1508).[61] In his history of Spain, the *Catálogo real* (1505–1532), Oviedo expresses his disapproval of Ludovico, referring to him as a tyrant and a usurper of the dukedom from its rightful heir, Gian Galeazzo Sforza (1469–1494).[62] Notwithstanding Oviedo's opposition to the duke's political maneuverings, in the four and a half months that he remained in Genoa he was able to appreciate Ludovico's financial support of the arts: since 1489, Leonardo da Vinci had been a salaried court painter at the ducal court of Milan.[63] Under Ludovico's patronage, he had produced a handful of paintings: *The Musician*, *Portrait of Cecilia Gallerani* (*Lady with an Ermine*), *Portrait of a Lady* (*The Belle Ferronnière*), a version of *Madonna of the Rocks*, and *The Last Supper*. Oviedo does not mention having seen any of these, but his high regard for the artist as an "excellent painter" indicates he was most likely acquainted with some of Leonardo's works.[64]

Of particular interest to Oviedo as a future illustrator of New World flora and fauna would have been the importance Leonardo ascribed to accurately reproducing the natural world in his art. An inventory the artist compiled of his portfolio in 1482 describes "many flowers copied from nature"; some of these sketches served as preparatory studies for paintings featuring plant life, such as *Ginevra de' Benci* (1474–1478), *Adoration of the Magi* (1481), and the *Madonna of the Carnation* (1480).[65] Leonardo's pictorial efforts to capture the true character of landscape and vegetation was not new, being preceded by Pisanello's (1395–1455) animal sketches and Netherlandish painting.[66] Nevertheless, Leonardo's treatment of nature was more scientific than that of any artist before him.[67] His detailed plant sketches anticipated the work of mid-sixteenth-century botanists who likewise documented such features, albeit for a different purpose. The degree to which Leonardo enlisted the natural world to inform his work helped legitimize its academic study.

The aesthetic criteria for which Leonardo advocated contributed to endowing his artistic subjects with life. In *Batallas*, Oviedo alludes to several of the hallmarks of a successful painter that he claims Leonardo had discussed in his presence—namely, an understanding of proportion as well as the presentation of the subject in an aesthetically appealing manner.[68] Above all, Leonardo stressed the need for relief, a technique he refers to as the "soul of painting" since tonal distinctions between light and dark provide art with its lifelike three-dimensionality.[69] For Leonardo, all of these strategies contributed to the ultimate objective of the artist: to communicate clearly and immediately with viewers. In his treatise on painting he writes,

> That science is most useful the results of which are most communicable, and, conversely, that is less useful which is less communicable. Painting makes its end result communicable to all the generations of the world, because it depends on the visual faculty; not reaching the understanding through the ear, it does not proceed in the same way that impressions of vision do. Thus painting does not have need of interpreters for different languages as does literature, and it at once satisfies mankind, no differently than do things produced by nature; and not mankind alone, but other living creatures also.[70]

Oviedo shared with Leonardo the belief in visual representation as an aid to understanding. In the *Historia general*, he explains his decision to include illustration alongside his written account via an argument reminiscent of Leonardo's opinions: "Without a doubt, the eyes contribute greatly to information about these things."[71]

In *Batallas*, Oviedo employs similar guidelines to assess a medallion owned by don Johan de Cardona. The Alcaide explains to Sereno that the pendant had been commissioned by the nobleman from Caradosso Foppa (ca. 1452–1526/27), a Milanese goldsmith. He recalls having held it in his hands, admiring its precious stones and observing the scene depicted: a knight, contemplating a very beautiful woman, evinces a profound sadness but is unable to cry.[72] Praising the artistry of the piece, the Alcaide proclaims it such an "excellent work that everything I have said could be understood by one with an average understanding."[73] For Oviedo, the medallion's value does not just reside in the material out of which it is made but also in its portrayal of the subject, which the viewer can immediately comprehend. He commends its maker too, alleging to have met Caradosso in Milan and calling him "one of the famous and fine men of his profession in that time in all of Italy."[74] Indeed, Caradosso had been employed by the Duke of Milan since 1475. Although nothing is known of his commissions in 1499, when Oviedo resided in the dukedom, his activities would have included cutting and mounting jewels and designing and creating decorative pieces such as Cardona's pendant.[75]

Mantegna and Martin Schongauer: Art at the Court of Mantua (1499–1500)

Oviedo's four-and-a-half-month sojourn in Genoa coincided with the end of Ludovico's rule over Milan. Sometime before September 1499, the duke fled the city before the approaching invasion of the king of France, Louis XII. In the following years, Caradosso and Leonardo left as well, the former to Rome and the latter to Mantua, the home of Isabella d'Este, Ludovico's sister-in-law. Isabella's court was Oviedo's destination as well; by the time Leonardo journeyed there, sometime during the winter of 1499–1500, Oviedo had likely already arrived. Much like her brother-in-law, Isabella was a patron of the arts and an intellectual in her own right. Since marrying the Marquis Francesco II Gonzaga in 1490, Isabella had made a concerted effort to make her presence felt in humanist circles in Italy: she was a prolific letter writer and a collector of antiquities, books and manuscripts, art, and musical instruments. Her *studiolo* in the Castello di San Giorgio of the Gonzaga palace held these possessions and served as an intellectual retreat.[76] Isabella worked to acquire paintings by the Renaissance artists Giovanni Bellini and Leonardo (the latter without success) and the sculptors Gian Cristoforo Romano and Pier Jacopo Alari Bonacolsi.[77]

One of Isabella's most fruitful artistic relationships was with Andrea Mantegna. The Paduan artist had been a court painter for her husband's family, the Gonzagas, since 1460, and he remained at Mantua until his death in 1506. By the time Oviedo met him, Mantegna had long been at the pinnacle of his career and was one of the most eulogized artists of his day.[78] The Neapolitan poet Jacobo Sannazaro, an acquaintance of Oviedo's, was one of his many admirers. He considered Mantegna "the shrewdest and most ingenious of all artists" and celebrated his talents in the pastoral novel *Arcadia* (1504), which Oviedo read.[79] When Oviedo arrived at the Mantuan court, the artist was busy finishing a major commission for the marquis, which he had begun fifteen years earlier, in 1484. Titled *The Triumphs of Caesar*, the work consisted of a series of nine nine-by-nine-foot canvases. The paintings were placed above eye level in a room in the Gonzaga palace for viewers to admire the painted figures, which seemed so alive as to breathe, and to appreciate the artist's technical skill.[80]

Although Oviedo does not mention having seen these paintings, nor any that Isabella d'Este had commissioned from the artist for her *studiolo*, he refers to Mantegna as "another Leonardo" and adds that some even regard him as the greatest painter in all of Italy.[81] In *Batallas*, he lauds a small, circular painting worn by a courtier named Diego de Quiñones that he attributes to the Paduan artist. Oviedo details the four scenes it depicts: in the first, a group of doctors and men of science debate among themselves; in the second, men play darts, cards, and chess; in the third, they eat and drink; and in the fourth, they engage in combat, on horseback and on foot.[82] Oviedo describes his impression of the painting as follows: "And all was executed with such perfection that it was marvelous to see the beauty and subtlety of the people and actions depicted in the painting, so much so that it was necessary that each be considered individually, and that he who looked at the painting have a knowledge of the art of painting to be able to enjoy its multitude of artistic achievements."[83] As in the case of the Caradosso medallion, here Oviedo admires the verisimilitude of the attitudes and actions of the subjects represented. He goes a step farther in his praise of Mantegna's painting, however. It is not enough that the viewer understands the scenes portrayed; he must also appreciate the artistry of their composition. This ability is not universal but rather restricted to those with more nuanced technical knowledge of the pictorial arts. It is clear that Oviedo considers himself well-informed on these matters, to the point that he is able to look beyond the mimetic nature of the work to contemplate the skills with which the artist constructed this artificial world. By participating in the humanist tradition of ekphrasis, Oviedo simultaneously pays homage to some of the most preeminent artists of his day—Mantegna and Caradosso—and demonstrates a familiarity with the technical standards of art in Renaissance Italy.

As much as Oviedo marveled at the artistic talent of these men, he was complimented by them in turn for his dexterity in making cutouts with paper and scissors. In *Batallas*, he recounts having performed this pastime in front of Prince John, the Duke of Milan, Isabella d'Este, and King Frederick of Naples, respectively, for their entertainment.[84] On one such occasion, the duke solicited Leonardo's opinion regarding Oviedo's paper creations. To his query, the Italian artist supposedly replied: "'Of all the things I have seen in the world, this is what has most amazed me, and if I myself had not seen him make these cutouts, I would not have believed that a man could do such subtle things with only scissors and without a drawing.'"[85] Oviedo claims to have similarly impressed Mantegna during his time in Mantua. Oviedo does not provide many details as to the form these creations took, but he explains that they are copies of "panels of Martinus and other great men of the burin."[86]

It is almost certainly the case that the "Martinus" to whom Oviedo refers is none other than Martin Schongauer, a German painter and the most accomplished engraver of the fifteenth century.[87] Although Schongauer died almost a decade before Oviedo's sojourn in Italy, in 1491, copies of his engravings were widely disseminated throughout Europe; as the first to mark plates with his initials, the prints Oviedo used would have been easily recognizable as the work of this German master.[88] Moreover, given the vogue in Spain for northern European art, it is likely that Oviedo had encountered his work there as well. Like Leonardo and Mantegna, Schongauer was praised by contemporaries for his close imitation of nature. In his engravings, he strove to accurately reproduce plants and animals and to create believable spaces through the introduction of volume and texture.[89] As Leonardo observes of Oviedo's method for making cutouts, he did not rely on templates; indeed, this is a practice that the Spanish historian criticizes.[90] Cutting these engravings from memory, therefore, necessitated Oviedo having studied prints by Schongauer and others carefully. Although artisans carved the woodcuts that Oviedo commissioned for the *Sumario* and the *Historia* (1535), he provided them with preliminary drawings. In executing sketches for this purpose, undoubtedly Oviedo employed his prior knowledge of the techniques of Renaissance engraving and woodcutting to inform his artistic decisions.

Alonso Berruguete: Art at the Court of Charles V (1519)

Taking his place among the handful of artists that Oviedo commends in his writings is one Spanish painter and sculptor. Alonso Berruguete exemplifies the awakening interest for Italian art in Spain at the turn of the sixteenth century. Son of the painter Pedro Berruguete (ca. 1450–1504), his first exposure to the Italian style of painting was likely imparted to him by his father, whose dominant Flemish aesthetic incorporated some Italian elements.[91] Alonso Berruguete's true education in Renaissance art, however, would not occur until after his father's death, during the ten years he resided in Italy (1506–1518). Although by 1515 he had begun to cultivate his reputation as a painter in Florence, Berruguete saw Charles V's ascension to the Spanish throne as an opportunity to gain power and prestige. He returned to his homeland sometime in 1518 to solicit, and eventually receive, an appointment as Painter to the King (*pintor del rey*).[92] It was here, in the royal court, where Oviedo became acquainted with the artist and his work.

It is possible that Oviedo met Berruguete for the first time in 1519; at the very least, and as we will see shortly, he was introduced to one of the artist's paintings that year. At the time, Berruguete was working on his first royal commission; the king had charged him with the decoration of the burial chapel of Jean Sauvage, his former grand chancellor, which consisted of a painted altarpiece and a sepulcher.[93] Oviedo applauds Berruguete's talents in sculpture and compares his works to those of the ancient Greek sculptors Phidias and Praxiteles and to King Mausolus's tomb.[94] However, it is unlikely that Oviedo had this initial commission in mind when lavishing his praise. There is no indication that Berruguete had received training as a sculptor in Italy, in which case in 1519 he would not yet have had the skills to undertake the carving of the sepulcher himself. Although in his official capacity Berruguete was expected to paint the altarpiece and to design Sauvage's tomb and oversee its completion, evidence suggests that the Castilian sculptor Vigarny executed the latter piece. Berruguete would not acquire a claim to fame as a sculptor of retablos until the mid-1520s and 1530s, around the time he finished the altarpieces for the chapel of the monastery of Nuestra Señora de la Mejorada (1523–1526) and the Monastery of San Benito in Valladolid (1526–1533). The scholar Richard Kagan hypothesizes that Oviedo saw Berruguete's famed retablo at San Benito, but he could have seen other works as well during his multiple visits to Spain in the 1530s and 1540s.[95] What is certain is that Oviedo followed Berruguete's career even after the Spanish artist withdrew from most of his royal duties, as he indicates Valladolid as Berruguete's place of residence.[96]

While Oviedo does not describe any particular plastic creation of Berruguete's, he does share his impressions of one of his paintings, a portrait of the Countess Ana de Cabrera. Through the character of the Alcaide, Oviedo recalls having been shown the portrait by her husband, don Fadrique Enríquez II, in Barcelona in 1519. The royal court moved from Zaragoza to Barcelona on January 24, 1519, allowing for the possibility that Berruguete painted the portrait there that same year. Oviedo greatly admires its execution: first, the pictorial likeness of the countess, whose spirit Berruguete captured so well that it startled Oviedo to see it. The second half of the composition is equally notable: Oviedo describes its subject as an eagle with her eaglets, some huddled around their mother in the nest and others cast out onto the ground. Those in the nest gaze upward, into the light of the sun, whereas the others look away. This scene illustrates conventional wisdom regarding the eagle, which, according to the accounts of Pliny and Isidore of Seville, abandons its offspring that are too weak to withstand the sun's rays.[97] By placing this mountaintop scene alongside the portrait of the countess, Berruguete equates her moral rectitude with the eagle's ability to face the sun directly.[98] Oviedo praises the verisimilitude with which Berruguete renders these figures, maintaining that it is "one of the most perfect paintings that the famous Berruguete painted in his entire life."[99]

Despite the differences between the artists Oviedo met in his lifetime, they all share notable similarities that speak to general conceptions of the artist and his art in Renaissance Italy. Key to the creative endeavors of Leonardo, Caradosso, Mantegna, Schongauer, and Berruguete was the desire to imitate nature and its processes. To achieve this objective, many of these artists turned their sights toward the world surrounding them. In the spirit of their Greek predecessors, they strove to produce works of art so closely resembling their subjects as to confuse reality with its representation. Oviedo's eulogies of these artists coincide with their reputation in society at large: celebrated in prose and verse as the most talented, skillful artists of their time, their art incarnated the values prized in the courtly circles that Oviedo frequented in Italy. The same standards Oviedo employs to describe and assess the paintings, medallions, and sculptures created by these men would play a major role in his approach to illustrating the natural world of the Indies in the *Sumario* and the *Historia general*.

Body and Soul: Medallions in *Batallas y Quinquagenas*

The interactions Oviedo sustains with the art and artists of the Italian Renaissance are only one aspect of his inquiries into the visual culture of early modern Europe. These experiences provided him with the knowledge to be conversant in the subject as well as a discriminating viewer. Oviedo's encounters with portrait medals offered him yet another opportunity to showcase this expertise and to put on display his multifaceted approach to graphic representation. In society at large, these medallions functioned simultaneously as celebrations of one's values and as decorative objects. Their bipartite purpose enables Oviedo to comment on the craftsmanship of a piece and its contents within a wider moral purview. In so doing, he contributes to the civic tradition of honoring the deeds of exemplary individuals chronicled by Pliny and revisited later by the Italian humanists, beginning with Petrarch. Oviedo's treatment of portrait medals in *Batallas* underscores how his understanding of visual imagery transcends aesthetic expression to encompass higher truths. Just as Oviedo's reflections on New World nature prompt him to pay tribute to the divine maker of the universe, so too do portrait medals invite him to participate in efforts to immortalize worthy men deserving of remembrance.

In the *Historia naturalis*, Pliny explains how statuary developed, beginning with depictions of gods and gradually giving way to statues of virtuous men. According to the Roman historian, the latter tradition began at Olympia, where those who had won the athletic competition three times were honored with statues modeled after their likenesses.[100] From this practice arose the custom of erecting "statues . . . as ornaments in the public places of municipal towns, and the memory of individuals was thus preserved, their various honours being inscribed on the pedestals, to be read there by posterity, and not on their tombs alone."[101] These statues were placed atop columns, to physically elevate these exemplary men above the rest of the populace. That their deeds be remembered was the primary aim of these statues for Pliny, ensuring that future generations had models to which they could aspire.

Centuries later, in Petrarch's *De remediis*, Reason laments the decline of statues from "hallmarks of virtue" to mere "attractions for the eyes."[102] This was, at least partially, Petrarch's own point of view. The Italian humanist understood glory or *fama* as the natural result of one's talents being used virtuously.[103] For this reason, he advocated for earthly immortality through the commemoration of personal accomplishments, as did Pliny before him.[104]

This parallel is particularly evident in the moralizing interpretation that Petrarch and others attached to Roman coins, one of the most abundant and easily collectable of classical antiquities. Their bipartite structure—with a portrait on one side and an allegorical or historical scene on the other—facilitated interpretations regarding the deeds, aspirations, intellect, and other virtues of the individuals depicted.

The popularity of these coins among early humanist circles contributed to the invention of the portrait medal in the fifteenth century.[105] Like their ancient counterparts, portrait medals (referred to in the Italian tradition as *impresa* and in French as *device*) consisted of two or three parts, in varying degrees: a portrait of the individual; another, usually allegorical, pictorial image; and a short motto or poem. Each of these components contributed to the piece's metaphorical meaning, which in turn spoke to the moral character of the subject of the medal.[106] As objects that circulated among friends and acquaintances and that were exhibited ostentatiously on one's person, portrait medals functioned as public declarations of one's identity. When documenting the lives of the Spanish elite in *Batallas*, Oviedo draws on these medallions as visible signs of the qualities and deeds these individuals prized above all others.[107]

Given that Oviedo began composing *Batallas* in 1535, a few words are in order as to the nature of his encounters with these medallions, which likely occurred at a much earlier date. It seems they were often kept on the person in question, either affixed to a chain and worn around the neck or stored in a pocket. Oviedo recalls many of them as being richly decorated with various gemstones, indicating that they had been made for display and were not meant to be distributed among a wider public. In lieu of such exchanges, it appears to have been common practice to circulate drawings of medals accompanied by their verses (many approximate the length of poems and could not possibly have fit onto the back of a coin, as the motto did). Indeed, both of Oviedo's alter egos, Sereno and the Alcaide, recount having received apparently unsolicited drawings of medals on various occasions.[108] It is possible that Oviedo had informed his acquaintances of his desire to document this information for posterity. Although most of these occurrences cannot be dated, Oviedo's interest in the portrait medal is apparent as early as 1499, when he met the goldsmith Cristoforo Caradosso Foppa, to whom he attributes the creation of don Johan de Cardona's portrait medal.[109]

At the same time as portrait medals provided Oviedo with the opportunity to learn about the personal accomplishments and aspirations of the Castilian elite, their inclusion in *Batallas* allows him to display his knowledge of art and its various levels of meaning. For Oviedo, a portrait medal's design is just as important as its content: the more elaborately decorated the medal, the more noble the individual who had commissioned the piece. So too, the originality of the image and its accompanying motto speak to the ingenuity of the noble in question, as well as the moral values he champions. The scenes commemorated on the medals drew inspiration from the natural world, daily life, the deeds of the subject portrayed, and allegorical motifs, principally amorous conflicts and hopes and the rises and falls of Fortune.

Don Johan de Cardona's medal of the knight gazing in sadness at an extraordinarily beautiful woman falls into the latter category. The motto chosen by don Johan to accompany the image speaks to a sadness so deep it cannot be conveyed visually: "Repressed tears / hurt more than those displayed."[110] Read in conjunction with one another, Oviedo assigns to the motto and the image the following meaning: "Don Johan . . . says that he will die without hope that others weep for him, nor that he himself can weep for himself or his pain."[111] His interpretation of the medal was likely influenced by don Johan's untimely end, wounded and captured after the Battle of Ravena (1512) as a defender of the Roman Catholic Church. Further contributing to Oviedo's understanding of don Johan's life and accomplishments is the fact that the nobleman sent several Petrarchan verses to a friend reflecting on his imprisonment: "None ever grieved for loss of liberty / Or doom of death as I for freedom grieve / and life prolong'd, who only ask to die."[112] The Alcaide explains to Sereno that these verses express don Johan's desire to have died in battle rather than live out his days as a prisoner. He depicts the suffering of don Johan as that of a man who served his country, his faith, and his people with the utmost virtue, even ordering that the ransom money given for his release be donated instead to the poor, as he was nearing the end of his life.[113] Oviedo's interpretation of don Johan's portrait medal at once pays homage to such an honorable man and praises the artistry of the piece.

Oviedo's treatment of portrait medals in *Batallas* indicates the extent to which art was integrated into the fabric of daily life among the elite in early modern Europe. Far from being a rarefied sphere of knowledge appreciated by the few who had access to the paintings and sculptures of artists like Leonardo and Mantegna, conversation about these topics abounded in the humanistic and courtly circles where Oviedo was present. The ekphrastic exercises involving portrait medals put into practice the urgings of Petrarch's Reason, who encourages Joy to look beyond the surface of a work of art to contemplate its deeper meaning. Medallions point to this higher truth by highlighting the virtues of their owners. For Oviedo, as for many others at the time, the moral aspect of a work of art—be it a painting, sculpture, or medal—was as important as its content and technical execution. Oviedo would embark on his explorations of New World flora and fauna with this frame of reference.

New Worlds, New Art

Over a decade before Oviedo set sail for the Indies for the first time, in 1513, he had experienced a new world of a different sort. The four years he spent in Italy (1499–1502) exposed him to the foremost artists of the period and provided him with the rhetorical conventions for commemorating these interactions in *Batallas*. From the rich dialogue that the Italian Renaissance established with its ancient past arose a shared understanding among Oviedo's contemporaries of "good" art as based on the imitation of reality and the measure of artistic genius as the ability to produce subjects so believable as to seem to pulse with life. To achieve these effects, painters and sculptors looked to the world around them as the ultimate model. Yet, despite their continual efforts to mimic nature's processes, there would always be limits to their ability to simulate reality. Don Hugo's laments on the imperfect pictorial representations of his beloved echo Oviedo's own understanding of the futility of any attempt to capture the likeness of the bird gifted to him by Captain Urdaneta. His views on the shortcomings of human artistry, however, do not prevent him from admiring the paintings, portrait medals, and sculptures he encountered in Italy and Spain, albeit with the caveat that the purpose of art should be moral edification. Nor do they deter him from attempting to replicate these efforts—albeit on a much smaller scale, and with the knowledge that he does not share the talents of these masters—in his writing on the Indies. Oviedo's debt to his Renaissance experiences is never so apparent as when reality manages to elude his efforts to capture it.

CHAPTER 3

An Island Kingdom

Flora and Fauna at the Service of Empire in the *Sumario* (1526)

The publication of the *Sumario* in 1526 inaugurated the literary genre of New World natural history. A testament to the work's success, it was immediately translated into Italian by the ambassador of the Venetian Republic, Andrea Navagero (1483–1529), and shortly thereafter published by the travel writer Giovanni Battista Ramusio as the *Libro secondo delle Indie Occidentali* (*Second Book of the Western Indies*) (1534).[1] Oviedo likely hoped more such recognition would be forthcoming. After all, at the time of the *Sumario*'s publication he had spent more than two decades assiduously documenting life around him with the aim of securing an appointment as court historian for the Spanish Crown.[2] These efforts first took shape in 1505, when Oviedo embarked on a genealogy of Spain's monarchs for Ferdinand, king of Aragon and recently appointed regent of Castile.[3] Unfortunately, Ferdinand's successor Charles V was more interested in establishing his legacy than in commemorating the history of the kingdom he had inherited, leaving Oviedo's manuscript, the *Catálogo real*, to languish for almost three decades.[4] It was only by turning his sights to the New World—a topic that did interest the young monarch—that Oviedo was able to come closer to fulfilling his career aspirations.

In the *Sumario*'s prologue, Oviedo explains that he wrote the work out of a desire to entertain Charles V with a novel subject, the natural world of the Indies.[5] By calling the emperor's attention to what Oviedo considers to be a neglected but crucial aspect of Spain's colonial enterprise, he aspires to secure Charles V's patronage for his next project, the ambitious *Historia general*. Oviedo excuses the brevity and disorder of the *Sumario* as the result of having written the work in Spain from memory, his notes in Santo Domingo.[6]

He assures the king repeatedly that the *Historia general* will be a more elaborate, well-organized account of New World nature, told alongside the history of its conquest and colonization.[7]

Yet Oviedo's claims in the *Sumario*'s closing paragraph that its successor was nearly finished, save a final "cleanup" of the manuscript, seem highly suspect.[8] Sections of the *Sumario* appear verbatim in the first part of the *Historia*, published in 1535, suggesting that much of the latter's composition occurred after 1526.[9] The close-knit relationship between the two works also indicates that the *Sumario* was a greater source of pride for Oviedo than what his protests of its flaws lead us to believe. At no point in his writing career does he repudiate the *Sumario* as he does his first publication, the chivalric romance *Claribalte* (Valencia, 1519). Oviedo does not mention the romance in any of his subsequent writings, and he is highly critical of this literary genre in the *Historia general*. In contrast, the *Sumario* forms the cornerstone of his lifelong project to chronicle the flora and fauna of the Indies.

Short and succinct, it comes as no surprise that the *Sumario* is frequently eclipsed by the monumental *Historia general*. Indeed, some even conflate the former with the latter, seeing little benefit in considering the work independently: Antonello Gerbi argues that there is "no real divergence of ideas between the minor [*Sumario*] and major work [*Historia*]" and Antonio Barrera-Osorio regards the two as a "single work."[10] Jesús Carrillo Castillo's *Naturaleza e imperio* (2004) and Kathleen Ann Myers's *Chronicle of America* (2007), perhaps the two most important contributions to our understanding of Oviedo's writings on the Indies in the last several decades, are both devoted to the *Historia general*.

Increasingly, however, scholars have begun to advocate for a renewed appreciation of the *Sumario*. Andrés Prieto laments its status as a mere "biographical curiosity," as does Jeremy Paden.[11] Taking a cue from Stephanie Merrim, Paden asserts that the publication of the *Sumario* marks the "initial systematic broaching of the problems of New Worlds and ancient texts" and sets in motion the continuous cycle of writing and editing New World natural and human history that would occupy Oviedo for the rest of his life.[12] Of particular interest to these scholars is the multifaceted role memory plays in the *Sumario* as an ordering tool and a persuasive device designed to advance Oviedo's career aspirations.[13] Still, a looming, unanswered question remains: Why did Oviedo decide to write a natural history in the first place?[14] Not only is the choice of topic unexpected, given Oviedo's previous work—the historical *Catálogo real* and the chivalric romance *Claribalte*—but its treatment is at odds with the portrayal of New World nature in the earlier accounts of Christopher Columbus, Hernando Cortés, and Pietro Martire d'Anghiera. These reports reduce flora and fauna to a secondary concern, interspersing observations of the natural world haphazardly among the main events of the discovery of the Indies. The initial glimpses of life on the other side of the globe that these sources provide us are a far cry from the systematic inquiry that defines Oviedo's approach to the subject.[15]

Oviedo's decision to foreground a topic of marginal importance to his predecessors speaks to the central role he assigns to the uses and value of nature in its myriad forms. Not only is he the first Spanish authority on New World nature but he is also its greatest champion. While the *Sumario*'s brevity may suggest otherwise, this is no easy project to undertake; ordering and naming the dense tropical wilderness of the Caribbean is a monumental task. The challenges Oviedo faces are twofold: on the one hand, nature is both the same and different in distinct territories, leaving Oviedo to struggle with how to address these similarities while avoiding redundancy. So too, despite the work's expansive claims to be a natural history of the Indies, Oviedo's reliance on his eyewitness testimony necessarily limits his knowledge of this topic to places he has been. In the *Sumario*, for the first time, Oviedo attempts to organize and communicate the lived realities of these overseas geographies to readers. The difficulties he seeks to resolve when doing so are the foundation of all his subsequent editorial interventions in this space.

Organizing Nature in the *Sumario*

Notwithstanding the novelty of its subject, Oviedo envisions the *Sumario* as part of an august tradition of documenting the natural world inaugurated by Aristotle (384–322 BCE), Theophrastus (371–287 BCE), and Pliny the Elder (23–79 CE). Of these classical models, it is the latter's encyclopedic *Historia naturalis* that most inspires Oviedo. In the *Sumario*'s prologue, he praises his Roman predecessor for being "better than any other author in that which concerns natural history."[16] Oviedo considers Pliny's work the result of a historian who "said what he could see and understand of such materials" but who also consulted and credited others when necessary.[17] Although Oviedo signals his intention to do the same, ultimately the only authority he cites in the *Sumario* is Pliny, and sparingly at that. Instead, Oviedo gathers information almost entirely through his own eyewitness observations.

In making the natural world of the Indies his principal subject of inquiry and himself the primary observer and authority on its wonders, Oviedo confronts various difficulties that others had not had to consider. First and foremost among them is how to impose order on this vast, chaotic wilderness. Present in the *Sumario* is the same hierarchy of animals, birds, reptiles, trees, plants, and precious metals that informs Pliny's work. Yet the need to differentiate between diverse New World ecosystems and territories prompts Oviedo to overlay this thematic organization with geographical divisions between the Greater Antilles (Hispaniola, Cuba, Jamaica, and Puerto Rico) and mainland Castilla del Oro.[18] In the case of Hispaniola, Oviedo provides an overview of the island under Spanish rule (1 chapter) and then describes its indigenous inhabitants (1 chapter), forms of sustenance (3 chapters), and birds (1 chapter). After briefly discussing Cuba, San Juan (Puerto Rico), and Jamaica in chapter 8, Oviedo moves on to the province of Castilla del Oro, which he organizes as follows: indigenous inhabitants (2 chapters), animals (17 chapters), birds (21 chapters), insects (5 chapters), snakes, lizards, and toads (8 chapters), flora (19 chapters), miscellanea (1 chapter), gold mines (1 chapter), fish (1 chapter), pearl fishing (1 chapter), and travel between the North Sea (Caribbean) and the South Sea (Pacific Ocean) (1 chapter).[19]

The stark disparity between the number of chapters Oviedo devotes to the islands versus the mainland, coupled with the overlapping existence of other, thematic divisions, has made the *Sumario* hard to define. Stephanie Merrim maintains that the work is essentially a miscellanea that "resists structuring."[20] In contrast, Carrillo Castillo argues that the *Sumario* obeys "a geographical logic that progressively dilutes into a thematic-encyclopedic organization."[21] The difficulties in determining the exact nature of the relationship between the two organizing principles at play in the *Sumario* reflect the struggles Oviedo faced in broaching such a multifaceted topic.

These problems are visible not only in the overarching structural tensions that Merrim and Carrillo Castillo observe in the work but also in the blurring of geographical divisions that pervades many of its chapters. Although Oviedo purports to devote chapters 2 through 8 to the islands and chapters 9 through 85 to the mainland, in truth he confuses these boundaries constantly, addressing island flora and fauna in his account of *tierra firme* and vice versa. So too, the mere seven chapters that Oviedo devotes to the archipelago in the *Sumario*, in contrast to the seventy-six on Castilla del Oro, beg the question of why he claims that the work is an advance on the *Historia general*. The first part of the *Historia* (1535)—and the one on which he was presumably working at the time of the *Sumario*'s publication—takes the islands, and specifically Hispaniola, as its point of departure, a geography that plays a minimal role in the *Sumario*. As I will argue in the following paragraphs, these features of the work speak to the realities of New World nature as well as Oviedo's biographical circumstances upon first arriving in the Indies.

In part, Oviedo's decision to privilege the natural world of the mainland in the *Sumario* is a response to the commonalities he observed between its ecosystem and that of the islands.[22] He mentions several species of animal that reside on Hispaniola—two types of rodent, the guinea pig and the hutia, as well as the iguana—only to explain that he will address them in the chapters on Castilla del Oro, "since they [these animals] are common among all of the Indies."[23] So too, he concludes his brief account of birds on Hispaniola, writing, "What can be said of the natural fruits of this land and its other plants and herbs and saltwater and freshwater fish is so little that I will not bother including it in this account of Hispaniola, because everything here is more abundant on *tierra firme*, along with many other things besides that will be said later on, where appropriate."[24] The division Oviedo imposes between the islands and mainland is one that, as he is aware, does not always correlate with natural patterns in the distribution of plant and animal life. Indeed, scientists have determined that most islands in the Caribbean archipelago display a variety of South and Central American plant species, in addition to their native flora.[25] Since, in Oviedo's view, the mainland hosts many of these species in greater numbers than its island counterparts, he reserves comment on them for the *tierra firme*. In so doing, Oviedo shows a preference for textual economy: to avoid repeating the same information relative to both the islands and the mainland, he determines to address certain phenomena in the geography where they are most prevalent.

That most of the chapters in the *Sumario* are dedicated to life on the mainland also attests to Oviedo's experiences in the Indies prior to 1526. Notably, Oviedo spent little time on Hispaniola or any of the neighboring islands in the first decade or so of his life in the Indies; he conducted most of his professional and personal life in Darién, in the province of Castilla del Oro.[26] It was here that, in 1513, the recently appointed governor Pedrarias Dávila was charged with taking the two thousand men who accompanied him on his voyage, Oviedo included. The expedition followed a route through the island chain of what today are the Lesser Antilles, bypassing Hispaniola entirely and arriving in Darién on June 30, 1514.[27] Oviedo lived at the settlement before his departure for Spain in 1515, and he built his first home overseas there upon his return five years later, in 1520. Oviedo reminisces fondly in the *Sumario* on the house, with its vistas overlooking the river, its accommodations fit for any great lord, and its orchard full of fruit.[28] It appears that Oviedo had intended to remain in Darién for some time: not only did he invest in the construction of a commodious home but his wife and children accompanied him on his voyage back from Spain in 1520.[29]

Oviedo mentions being on Hispaniola only three times prior to the publication of the *Sumario*, for the purpose of travel to and from Spain. From statements Oviedo makes in the *Historia general* regarding these voyages, he spent a total of roughly two months there before 1526, as compared to the three and a half years he lived on the mainland.[30] The journey from Darién to Santo Domingo (or, alternatively, from Cuba, where he made a stopover en route to Hispaniola in 1523) was not a direct one: Oviedo disembarked on the western half of the island, at Yaguana, and from there traveled eastward by land for eighty leagues, or roughly three hundred miles, to Santo Domingo.[31] In his traversal of the island, Oviedo would have seen much of its geography and come into contact with some of its flora and fauna. It is reasonable to assume, however, that he was more familiar with the natural world of the mainland, given his sustained personal and professional ties to the area.

Oviedo did not move to Hispaniola until immediately before his second trip to Spain, in 1523—just in time, too, as discontented natives burned his home in Darién to the ground the following year.[32] Barring a two-year sojourn in Nicaragua from 1527 to 1529, Santo Domingo remained Oviedo's home for the rest of his life; its pivotal role as a port from which all mainland expeditions were supplied and launched enabled him to stay up-to-date on the latest news relating to the discovery, conquest, and colonization. So closely do we associate Oviedo's task as a historian of the Indies with his life in Santo Domingo that it is easy to forget his initial years in the New World were not spent there. Still, Oviedo's residence in Darién is fundamental to the decisions he makes in the *Sumario*. This publication represents his first attempt to produce a natural history of the New World that is thorough but not repetitive and which privileges his lived experiences. To do so, Oviedo must reconcile the geographical divisions of the work with practical realities and editorial concerns. These are the same challenges with which Oviedo will wrestle throughout his writing career.

Forsaken Land: The Fight for New World Nature and the *Sumario*

To a large extent, Oviedo attributes the novelty of the *Sumario* to his knowledge of New World nature, which is superior to that which others have. This strategy is comparable to the one Oviedo will use later, in the *Historia* (1535), where he sets himself apart from "other writers"—specifically the unnamed but ever-present court chronicler Pietro Martire d'Anghiera—who write about life overseas without ever stepping foot there.[33] Yet in the *Sumario* these "others" assume a much broader profile: Oviedo targets not only those who have written on the natural world of the Indies from "here" (that is, Spain) but also other eyewitnesses to its wonders.[34] Regardless of the truthfulness of their accounts, Oviedo claims that since "they come to these parts to do business or work in areas that are of greater interest to them . . . they forget about these things, because with less attention they look at and consider them than does one who, naturally inclined [to these things] as I am, has wished to know them."[35]

Scholars have interpreted this passage to mean that Oviedo's stance toward New World nature is more objective than that of his compatriots, concerned first and foremost with financial gain. Carrillo Castillo conceives of Oviedo's approach to his task as "altruistic distancing, removed from an interested and commercial vision . . . the result of his 'natural inclination,' an innate inclination, more than an acquired habit or from a position related with the performing of his official duties."[36] Similarly, Paden argues that Oviedo sets out to conduct a "disinterested and purposeful study of these new lands . . . [not] impeded by pecuniary interests."[37] Oviedo does indeed distinguish between those travelers whose insights concerning the natural world are acquired incidentally, in the pursuit of other aims, and himself, for whom the flora and fauna of the Indies are principal subjects of inquiry. Far from casting himself as a distant and disinterested spectator of nature, however, Oviedo states explicitly that the natural world *is* his ultimate concern. That is to say, he trains the same attention on New World plants and animals that others dedicate to their business ventures. Oviedo makes no avowal of objectivity but rather of rigor and comprehensiveness.

The *Sumario* embodies a highly mediated vision of the natural world of the Indies. It is abundantly clear from Oviedo's allusions to the as-yet-incomplete *Historia general* that he had his own self-serving agenda in writing the work. According to Stephanie Merrim, in promoting his ongoing historiographical project in the *Sumario*, Oviedo "makes knowledge his commodity, his stepping stone to power and *fama* (fame).... The 'interested' and self-interested writer has effectively turned knowledge into a valuable unit of exchange."[38] But it would be imprudent to maintain that the work is so one-dimensional as to serve merely as an attempt to fulfill Oviedo's career aspirations. Instead, and just as importantly, the *Sumario* also registers his concerns about the future economic well-being of Hispaniola. Oviedo's faith in the commercial possibilities of the island is coupled with dismay at what he has witnessed of more popular methods of generating profit from its soil, which depend almost exclusively on the pillaging of its mineral resources. In the *Sumario*, Oviedo counters what he views as the shortsightedness of these actions by engaging in a vociferous defense of nature's nearly infinite potential as an overlooked and neglected source of utility and revenue.

Prior scholarship has confirmed Oviedo's opinions regarding "good" governance of the Indies as based on commerce, the cultivation of the land, and peaceable rule. As Antonello Gerbi articulates succinctly, "Oviedo . . . puts the colonist before the conquistador, and the farmer before the treasure-hunter."[39] His proposal for the governorship of Santa Marta, which he presented to the king in 1518, is often taken to epitomize Oviedo's ideal model of colonization, pitting his imagined nonviolent colonial society against the brutality and ransacking of those colonial officials of whom he is most censorious.[40] Ultimately, Oviedo's plan was never put into practice: the crux of the proposal—a demand for one hundred knights of the Order of St. James of the Sword to administer the *encomienda* system—was not approved, causing him to reject the appointment.[41] In the *Historia general*, however, Oviedo continues to insist on the arrangement as the solution to most effectively "govern and populate the land in the shortest amount of time."[42] Oviedo does not recount this event until several decades after it took place, in the Monserrat manuscript (1535–1549). That it remained unmentioned in the *Sumario* and the *Historia* (1535) perhaps accounts for the lack of scholarship on how Oviedo's vision of New World settlement informed these publications, and particularly the *Sumario*, the work that followed most closely on the heels of Oviedo's disappointments at court. In what remains of this chapter, I examine the *Sumario* for what it reveals about Oviedo's understanding of the role of the colonist and the Crown in enabling the survival and success of Spain's overseas colonies.

The close-knit relationship Oviedo establishes in the *Sumario* between commerce, empire, and empirical practices is a prime example of how certain individuals sought to benefit from the knowledge they accrued about New World nature in the initial decades of the sixteenth century. Antonio Barrera-Osorio has demonstrated how a host of actors—entrepreneurs, merchants, and bureaucrats—submitted natural phenomena to observation and testing in the hope of acquiring commercial license from the Crown to exploit them. Over time, these independent initiatives gave rise to an institutional culture of empiricism, as the Crown, searching for new sources of profit, devised mechanisms for evaluating the merits of the proposals it received.[43] There is a noticeable omission among those whom Barrera-Osorio identifies as participating in this burgeoning scientific culture: in his estimation, natural historians and physicians remained excluded from this community because they viewed natural phenomena as sources of "wonder" rather than as having material value.[44] Yet these two perspectives are not mutually exclusive; Columbus's profusions of admiration go hand in hand with his certainty in the economic promise of these new lands. In the *Sumario*, Oviedo too depicts nature simultaneously as a source of beauty and economic benefit.

The idea that Oviedo advocates for the material advantages of the natural world of the Indies is at odds with the prevailing understanding of the *Sumario* as divorced from these concerns. Recall that Carrillo Castillo and Paden maintain that Oviedo's stance toward New World nature is not impacted by commercial considerations.[45] Similarly, Merrim argues that "for Oviedo the real currency of the *Sumario* lies in a less material commodity. . . . Only at the end of his work, in fact almost as an afterthought, does its author directly address the economic advantages of the new territories."[46] The reason Oviedo defers a discussion of gold mining and pearl fishing until the *Sumario*'s last chapters, however, has little to do with his supposed disinterest in nature's economic possibilities. Oviedo was simply following the example of his revered predecessor: Pliny does not address precious metals until the final books of the *Historia naturalis* (*Book 36* and *Book 37*). Equally concerning is the assumption that Oviedo viewed mineral wealth as the only means to acquire material advantage. Certainly, Oviedo recognizes the benefits to be had from gold and pearls, but he also sees value in other aspects of New World nature that his compatriots have overlooked. The *Sumario* celebrates the New World for the untold advantages it will provide to Spain, so long as colonists are able and willing to harness its possibilities.

A Cautionary Tale: The Crisis of Hispaniola's Economy (1514–1520)

In the decade or two after Columbus landed on the shores of Hispaniola on December 6, 1492, the idyllic image he had painted of the island in his letter to Luis de Santángel (1493) gave way to the dire reality. In 1514, the same year as Oviedo's inaugural voyage to the New World, residents on the island were confronted with news of the drastically declining indigenous population: the census, conducted to redistribute the *repartimientos* that allocated native labor among colonists, records a total of 26,334 natives (including those who had been brought in from neighboring islands to replace the dwindling labor supply). This number is a far cry from the 60,000 reported on the 1508 census, which itself likely represents only a small fraction of the indigenous population living on Hispaniola prior to European contact.[47] The impact of this decline on the island's workforce was only one of several factors that jeopardized its economic well-being. Gold placers were drying up, and residents left to search for fortune elsewhere. When Diego Velázquez sailed for Cuba in 1511, he was followed by so many from Hispaniola that the western villas were eventually nearly emptied of inhabitants.[48] This exodus resulted in orders being issued to Pedrarias to bypass Hispaniola entirely on his 1514 expedition to Darién, to avoid recruiting men from the island.[49]

Oviedo would have witnessed the effects of this economic collapse in 1515, during his trek across the island on his way to Santo Domingo and, ultimately, to Spain, where, as *procurador*, he planned to lodge complaints against Pedrarias's governorship of Castilla del Oro. One of the fifteen villas on the island had been abandoned by that year, and others were in varying stages of neglect or in the process of being consolidated with other villas.[50] Oviedo mentions in the *Historia general* that he had been in Santa María de la Vera Paz in 1515 shortly before it was relocated to the harbor of Yaguana (present-day Léogâne) and renamed Santa María del Puerto.[51] The environmental impact of the island's dwindling population would not have been lost on him as he passed through: footpaths overgrown with weeds and second-growth forests, and cultivation plots (*conuco*) overrun by shrubs and other vegetation.[52]

So acute was the crisis that the island's residents began to press the Spanish Crown for a solution: in July 1515, the bishop of Concepción de la Vega, don Pedro Zuarez de Deza, wrote a letter to King Ferdinand, expressing his unease over the depopulation of the island and suggesting potential remedies.[53] In another letter dated October 6, 1515, officials on Hispaniola excused the lack of gold that had been sent to Spain as a consequence of the reduced native workforce. At the same time, they mentioned other natural resources of use and possible profit on the island, among them one that they sent to Spain for testing, the fruit of a tree they called *cañafístola* (*Cassia fistula*).[54] Not content to rely on written correspondence alone, several months earlier, in August 1515, these officials had appointed Gil González Dávila, an inhabitant of Santo Domingo and *contador del rey*, to offer an account of the island's affairs to King Ferdinand.[55] It is possible that Dávila and Oviedo were passengers on the same ship: both are mentioned in the letter penned in October 1515 regarding the official commissions they had been given (Oviedo as *procurador* of *tierra firme*), and both journeyed from Santo Domingo to Spain in the latter months of 1515.[56] Unfortunately, upon arrival their grievances remained unaddressed: neither was granted an audience with the ailing king before his death on January 23, 1516.

In the memorials Dávila wrote in preparation for his audience with King Ferdinand, which came into the possession of Cardinal Francisco Jiménez de Cisneros (1436–1517) following the monarch's passing, the *contador del rey* seeks answers for the island's population decline.[57] He attributes the problem to four causes, including the changing of governors; the continual reassignment of the natives in *repartimientos*; and legal conflicts between residents.[58] But it was the fourth cause that Dávila identified as having the greatest impact on the natives—namely, that "the objective has always been that as much gold as possible come from the island."[59] Among the remedies Dávila suggests are measures to diversify the economy by shifting resources away from gold mining to agricultural production.[60] He recommends that residents on Hispaniola develop other sources of income through the planting and harvesting of cotton, sugar, *cañafístola*, wheat, and grapes, and that laborers be sent from Spain for this purpose.[61] In the *Sumario*, Oviedo would advocate for some of these same measures as a means of increasing the economic potential of the island.

Dávila's memorials reiterated concerns voiced by members of the Dominican order, who had long advocated for political and economic reform in the Indies.[62] Heeding these growing outcries, Cisneros, acting as regent of Castile (January 23, 1516–November 8, 1517), commissioned the jurist Juan López de Palacios Rubios and Bartolomé de las Casas to propose solutions to the problems on the island (in the end this task fell primarily to Las Casas and his collaborator, fray Antonio de Montesinos).[63] Cisneros also instructed three Hieronymite monks to travel to Santo Domingo and evaluate the situation firsthand. He awarded them a joint appointment to the governorship of Hispaniola and gave them leave to enact any measures they found necessary to improve conditions there. Finally, Cisneros appointed *licenciado* Alonso de Zuazo as *juez de residencia* to conduct reviews of the judicial authorities on the islands and mainland responsible for managing the properties and revenues of the Crown.[64]

Independently of one another, the Hieronymites and Zuazo issued a series of recommendations to the Crown to inject new life into Hispaniola. In substance, these echoed the conclusions that the bishop of Concepción and Dávila had reached several years earlier: with gold production in perilous decline, the native population dwindling, and residents abandoning the island, all agreed on the pressing need to exploit Hispaniola's other natural resources. In contrast to the reports of the bishop and Dávila, however, the prevailing tone in these later accounts is one of increasing apprehension. "Truly, it is not a land to forsake," plead the monks, assuring Cisneros that if their advice is followed "it will be the best and the richest land, with the most income."[65] They recommend the planting and cultivation of wheat, grapes, sugarcane, cotton, and *cañafístola*. Given the little that colonists know of agricultural practices on the island, this will not be an easy task; the monks acknowledge that "the land is not cultivated and it is not known when the seeds should be sown and the plants planted." Nevertheless, they express confidence that "with effort much of this knowledge can be attained with the help of God."[66]

The *licenciado* too portrays Hispaniola as "the best land in the world."[67] He expands on the list of crops mentioned in the Hieronymites' report, highlighting not only the commercial possibilities of nonnative species but also of the island's indigenous flora. Zuazo mentions only one of these autochthonous species by name—the *brasil guayacan*, whose wood was already being used to treat illnesses associated with leprosy. But he is optimistic about the existence of other trees that could provide medicine and spices as well, such as a tree bark he identifies as having a taste similar to that of cinnamon (*Canella alba*).[68] Notably, Zuazo shares both the Hieronymites' conviction of the economic potential of the land as well as the recognition of the need for more extensive knowledge of it to better capitalize on these opportunities. He is confident, however, that with adequate resources and support "all will be understood with great diligence."[69]

The Hieronymites' governorship of the island (1516–1519) coincided with the dates of Oviedo's first trip back to Spain from the Indies (1515–1520). He followed news from Hispaniola with interest and a bit of annoyance, attributing his difficulties in obtaining an audience with Cisneros to his being wholly occupied with the Hieronymites' enterprise.[70] Oviedo speaks favorably of the monks' and Zuazo's accomplishments in the *Historia* (1535), lauding them for having invested in the island's infrastructure and in the construction of sugar mills.[71] At the same time, he criticizes their decision to consolidate the natives into villages, writing that "changing habits, especially among the natives, is to cut off their life."[72] The project was doomed to failure even before it began: a smallpox epidemic devastated the island, killing a third of its indigenous population.[73]

The monks' rule of Hispaniola was the first and last attempt the Crown made to intervene in the island's problems. The growth of the sugar industry allowed residents to transfer their fortunes from mines to mills, but no other large-scale efforts were made to diversify the island's economy and exploit its other natural resources.[74] The Crown's attention quickly shifted elsewhere, as Charles V became occupied with news of the marvelous wealth of the Aztec Empire and struggles to suppress the uprising in Castile that would later be known as the Revolt of the *Comuneros* (1520–1521). In contrast, the *Sumario* (1526) focuses anew on the Caribbean. Underlying Oviedo's explorations of flora and fauna in this work are the same concerns that had found Cisneros's sympathetic ear a decade earlier—namely, the need to learn more about New World nature to support the continuing needs of a colonist population. The vision of the natural world that Oviedo presents is one of nearly limitless potential. Many of the plants and animals that Oviedo describes not only provide the material conditions of life for those overseas but are also untapped sources of wealth for the Spanish Crown. In the *Sumario*, Oviedo advocates for a multifaceted approach to the Indies that acknowledges the diversity of its resources and the need to take advantage of them.

Knowledge and Advocacy: Defending Nature in the *Sumario*

Writing the *Sumario* several years after the failed attempts at reform on Hispaniola, Oviedo continues to observe the same deep-seated problems that Zuazo and other defenders of the island had identified. Indeed, these issues are at the heart of the work: articulated in chapter 2, effectively they inform the descriptions of flora and fauna that follow. To begin, Oviedo launches a twofold complaint regarding the disappearance of natives and settlers alike from Hispaniola: "This island was once very populated by natives. . . . At present there are very few and the Christians not as many as there should be owing to the many that have moved on to other islands and *tierra firme*. Those who come here, in addition to being friends of novelty . . . because there have been discovered and every day they discover new lands, it seems to them that they will fill up their pockets more quickly [by going to these places]."[75] Oviedo laments the widespread attitude toward the New World as a source of rapid fortune rather than a place of settlement. So pervasive is this desire, he adds, that many have foolishly abandoned homes and properties on Hispaniola to pursue delusions of wealth.[76]

Not only are these riches elusive but, in casting their gaze to distant shores, these fortune hunters ignore the treasures surrounding them. Oviedo champions Hispaniola as having untold value, echoing the sentiments of the Hieronymites and Zuazo. In his praise of the island and its resources, however, Oviedo's enthusiasm is beyond compare. "Without a doubt," he affirms, "if a prince did not have any larger an estate than this island alone, in little time he would outrival Sicily and England, nor at present are either of them to be envied. For what is superfluous on this island could make many provinces and kingdoms wealthy."[77] Ignoring the severe depletion of gold on the island, Oviedo alleges that Hispaniola has the "richest mines of the best gold that until now has been discovered in such quantities in any part of the world."[78] Its mineral wealth is just one of the natural resources the island has to offer. Oviedo boasts of the large number of nonnative species it hosts, which grow there "much better, and in larger quantities than anywhere in Europe": sugarcane, sweet and bitter oranges, figs, lemons, date palms, citrons, pomegranates, an assortment of vegetables, and other plants and trees.[79] Oviedo also points to the growing industry of livestock production on the island—the sheep, pigs, and cows that multiply at dizzying speeds.[80] Truly, for Oviedo, Hispaniola is a kind of Eden, its fertile soil and tropical climate allowing crops and animals to thrive more successfully than in their native land.

Reiterating earlier calls to diversify the island's economy, Oviedo names a handful of nonnative crops that have become commercially profitable, including wheat, grapes, sugarcane, and cotton, and others that have similar potential. Of the sugar produced on the island, Oviedo reports that it is of such good quality and so abundant that "ships load themselves up with it each year."[81] The island's indigenous *cañafístola*, too, is "brought to Spain in large quantities and from there it is taken to many parts of the world."[82] Yet many plants remain underutilized; Oviedo affirms that Hispaniola's wild cotton would be better than any in the world—ostensibly even better than that brought from Spain and grown on the island—were it to be cultivated and treated.[83] He attributes the cause of such neglect to the same problems he mentions at the beginning of the chapter—namely, that those who come to the Indies are greedy and not willing to labor the land. "For this reason," Oviedo writes, "they cannot be bothered to make bread or put in vineyards, because in the time that these things take to yield fruit, they find them for good prices [elsewhere] and ships bring them from Spain, and working in the mines or dealing in goods or in the harvesting of pearls or in other exercises, as I have said, they amass fortune more quickly."[84] Oviedo does not deny the enormous benefits to the Spanish Crown of the gold found on the island and elsewhere in the Indies. At the same time, however, he maintains that the colonization of the New World necessitates a more balanced economic model that encourages settlement, not plunder. Although some long-term colonists on Hispaniola grow a variety of crops, Oviedo identifies the majority of those who come to the Indies as young men, unmarried and unwilling to take up residence there permanently.[85]

Effectively, these comments serve as a preamble for the rest of the work. Though in chapter 2 Oviedo remarks on the successes that nonnative plants have experienced in the Indies, he reserves the majority of the *Sumario* for showcasing the potential of the New World's autochthonous species. By enumerating in one chapter after another what he knows of these phenomena, Oviedo confirms the medicinal, alimentary, and commercial possibilities of the flora and fauna of the Indies. The *Sumario* is a handbook for the kind of lifestyle that Oviedo seeks to promote overseas. He urges readers to look beyond the more obvious sources of wealth, which oftentimes elude even the most enterprising of men, to cast their eyes on the possibilities of nature at large.

In advocating for the benefits and profit to be had from New World plants and animals, Oviedo contributes to previous efforts to remedy the dire situation on Hispaniola. Be they fruits, grains, medicinal herbs, fish, birds, or land animals, Oviedo fervently believes, as did many at the time, that the natural world exists to serve mankind. In chapter 2 of the *Sumario*, Oviedo awakens readers to the nearly infinite possibilities of the land and its products in the Indies. Despite the striking similarities between the proposals of the bishop, Dávila, the monks, and Zuazo, and the ideas that Oviedo espouses, just as important are their differences. Let us recall the monks who, in their report to Cisneros, recognize their limited knowledge of the planting and harvesting of crops on Hispaniola. Then there is Zuazo, optimistic as to the botanical potential of the Indies but ignorant of this reality. In the *Sumario*, Oviedo does more than share in the vision of his compatriots; he supplements it. Putting on display his knowledge of the plants and animals of the Caribbean, Oviedo demonstrates that he is the expert on New World nature. This is not just a rhetorical gesture but one with real, material consequences; Oviedo holds an understanding of New World nature to be key for the success of Spain's colonizing venture.

Encomiums of Nature in the *Sumario* and the *Historia* (1535)

The *Sumario* is the foundation of Oviedo's subsequent attempts to render New World nature intelligible. As such, it comes as no surprise that the same concerns he expresses in this earlier work reappear ten years later, in the *Historia* (1535). His defense of Hispaniola assumes even greater proportions in this later publication, where he devotes an entire chapter to celebrating Hispaniola's resources and condemning those who set their sights on plunder rather than settlement.[86] If Oviedo had been worried about the future of the island in the *Sumario*, he had even more cause for concern by the 1530s, when trans-Atlantic trade with the Caribbean islands declined precipitously. Prior to 1530, these routes had accounted for 70 percent of commerce between Spain and its colonies; after, they represented a little over 30 percent. The alarm bells that Oviedo and his predecessors had sounded came to naught, as attention shifted to those regions with greater opportunities for enrichment, such as New Spain and Peru.[87]

In addition to reiterating his earlier sentiments from the *Sumario* regarding the situation on Hispaniola, in the prologue to the *Historia* (1535) Oviedo reflects on the state of affairs in the Indies at large. His prose becomes almost lyrical as he considers the cornucopia of resources and peoples that comprise Spain's overseas territories: "What mortal wits could comprehend such diversity of languages, habits, and customs, as that of the people of these Indies? Or such a variety of animals, both domestic, savage, and wild? Or such a multitude of trees, so overladen with diverse fruits, and others infertile. . . . So many plants and herbs of benefit for man. So many innumerable others that are not known and that are so different in their flowers and fragrances."[88] Oviedo mentions birds, meadows, fields favorable for agriculture, mountains, forests, beaches, rivers, lakes, fish, gold, silver, copper, and pearls in the same panegyrical tones. Nonnative plants and animals thrive in these climates as well: Oviedo remarks on the explosive growth of livestock, as well as the vegetables, fruits, and other products brought from Spain, which this land "has received not as a stepmother but even more like a true mother than that which sent them forth, for some of them are produced in greater quantities, and of a better kind, than in Spain itself, as well as animals . . . and bread and vegetables and fruits and sugar."[89] The New World is second to none in the variety and abundance of its nature world, as well as its embrace of new plants and animals.

Unfortunately, Oviedo regrets, little is done to take advantage of this enormous potential. The pursuit of gold, silver, and pearls leaves merchants largely indifferent to the commercial possibilities of other natural resources. Yet the few exceptions to this tendency—sugar and *cañafístola*, which are shipped to Spain by the boatful to sell—serve to demonstrate the financial promise of botanical species.[90] Oviedo urges similar attention to other possible sources of profit, arguing, "They could do the same [as they do with sugar, etc.] with other things that here are forgotten and that these Indies produced before the Spanish discovered them and that they continue to produce."[91] Orchella weeds (*orchilla*), brazilwood, and alum (*alumbre*) take their place alongside the cotton and *cañafístola* that Oviedo had mentioned in the *Sumario* as products that other kingdoms desire.[92] The nine years that have elapsed since the work's publication have not been without consequence: Oviedo has acquired further knowledge of New World phenomena. Nevertheless, the need to capitalize on these opportunities remains just as pressing a concern as always.

Knowledge of New World nature is nowhere more vital than in the day-to-day life of the colonies, where it yields innumerable forms of sustenance, remedies, and tools for its inhabitants. For Oviedo, expertise in this subject is a political imperative as well: wielded authoritatively, such knowledge enables effective government.[93] The anecdote Oviedo recounts in the *Historia* (1535) about the peace brokered by Captain Francisco de Barrionuevo with the Taíno cacique Enrique on Hispaniola demonstrates how victory can hinge on familiarity with one's surroundings. "No manner of food do we find in these parts that is like that which our fathers gave us," Oviedo informs his readers,

> The bread is made of roots; the fruits are wild and unknown [to us] nor do they agree with our stomachs; the waters taste differently; there

> is no meat to be found on this island except for those mute dogs that I have described and a few other animals different from those in Spain. Serpents like iguanas, snakes, and lizards. . . . And since the captain had tried all of these things on the island of San Juan and in the land to the north in Florida and in other parts he knew how to achieve his ends [with don Enrique]. Without a doubt, I believe that if he had been one who had just arrived from Spain, this peace would have never been accomplished.[94]

Oviedo attributes Barrionuevo's success in resolving the conflict with Enrique not to his military capabilities, as might be expected, but to his insights into New World nature. This knowledge, acquired by a painful process of adaptation to entirely new circumstances, is a political necessity: so central is this lived experience to the outcome of the captain's mission that Oviedo maintains that one lacking it would fail where Barrionuevo triumphed. In casting the captain's actions in this light, Oviedo demonstrates the invaluable wisdom of longtime residents in the Indies. The *Historia* (1535), a product of many of the same struggles and difficulties as Barrionuevo experiences, provides readers a glimpse of these realities, so vital to Spain's colonial endeavors.

In short, Oviedo's defense of nature as an overlooked and neglected source of utility and revenue for the Spanish Crown in the *Sumario* is the cornerstone of his treatment of this topic at large. At the heart of Oviedo's writings on the Indies is the belief that Spain will only achieve complete political dominion and economic prosperity in the New World once the great potential of its flora and fauna is acknowledged and harnessed. The *Sumario* represents the first of many attempts, in the sixteenth century and beyond, to apprehend these vast reaches of space and put them at the service of empire.

CHAPTER 4

The World as One and Many

Constructing New World Nature in the *Historia* (1535)

In the nine years that elapsed between the publication of the *Sumario* and that of the first part of the *Historia general*, in 1535, Oviedo became increasingly aware of the difficulties he faced in documenting New World nature.[1] These challenges are put on display most clearly in the prologue to *Book 9: On Non-Fruit Bearing Trees*, where he writes:

> The land is so covered with trees with so many differences among them, in the size of the trunk, branches, and leaves, as well as in their appearance and fruit, that not even the natives know them all or can give them names. . . . In many parts one cannot see the sky from below these groves of trees, since they are so tall and thick and full of branches, nor walk among them, because . . . there are other plants and greenery so intertwined and jumbled together with so many spines and vines and other branches that it is only with much work and the aid of knives and axes that a path can be opened. And what can be said in this regard is that it is a *mare magno* [great sea] and hidden, because most of what is seen is unknown, since men do not know these trees nor their properties.
>
> Some of them [trees] have fruit, others are in bloom, and others are beginning to bloom. And since they are all different kinds, they enjoy the seasons in different ways. And everything is seen together at any time of year; therefore, I will conclude here, because only with time will these infinite kinds and manifold differences in plants be understood.[2]

Acutely aware of the insufficiency of his efforts, nevertheless Oviedo strives to chronicle the flora surrounding him as best he can. This is no small undertaking: to name and describe these New World species, Oviedo must untangle and hack through—both literally and metaphorically—impenetrable tropical rainforests to bring some semblance of order and meaning to what he sees. The final product of these endeavors—the published *Historia* (1535)—is a testament to Oviedo's resolve in making known these vast reaches of wilderness.

The *Historia* (1535) is the last of Oviedo's writings on the Indies to see its way into print in his lifetime. It is, moreover, the work that cemented his reputation as a New World historian, both among his contemporaries and in the following centuries. Unique among early accounts of the discovery and conquest of the Americas for its encyclopedic, systematic treatment of flora and fauna, this publication marks the first fully developed articulation of New World natural history. With its emphasis on empiricism and eyewitness testimony, the work offered a model for subsequent natural histories and served as an indispensable reference for historians, scientists, and travelers throughout the colonial period.[3] By the mid-sixteenth century, a reprint had been issued (Salamanca, 1547), and French and Italian translations made possible its circulation among a wider readership.[4] The Italian travel writer Giovanni Battista Ramusio published a version of the work in Italian in volume 3 of *Delle navigationi et viaggi* (*Voyages and Travels*) in 1556, accompanied by a reprint of the Italian translation of the *Sumario* (Venice, 1534). In the same year, in Paris, Michel de Vascosan published Jean Poleur's French translation of the *Historia* (1535). These Italian and French editions feature Oviedo's woodcuts as well (some of them substantially reworked), presenting readers with a twofold verbal and visual description of New World flora and fauna akin to that of the original Spanish edition.

The *Historia* (1535) is the definitive work on New World nature in the first half of the sixteenth century, and Oviedo its greatest authority. By positioning an account of the plants and animals of the Indies alongside the narrative of its discovery, Oviedo advocates for nature as a subject just as deserving of attention as its human counterpart. To give literary form to this all-encompassing reality, Oviedo must work within his limited means to bring the subject to life; these efforts are at the heart of his construction of New World nature in the *Historia* (1535).

The *Sumario* and the *Historia* (1535)

Nearly ten years after Oviedo announced in the *Sumario* his intentions to write a "general and natural history of the Indies," the long-anticipated first edition of the *Historia* came off Juan Cromberger's press in Seville.[5] In the prologue, Oviedo describes the work as elaborating on its predecessor's account of the natural world of the Indies, in addition to including a history of its conquest and colonization.[6] Oviedo's insistence that this information will be "better and more copiously said" than in the *Sumario*, however, does not begin to do justice to the *Historia*'s monumental proportions.[7] Consisting of twenty books, the 1535 tome is not only much bigger than the earlier work but also contains nearly seven times more woodcuts. At the time of its publication, the project was not nearly complete: Oviedo explains that he envisioned the as-yet-unfinished *Historia general* as embracing the present volume as well as two future installments.[8]

Ultimately, the final product exceeded even these immodest expectations. In 1549, when Oviedo deposited the so-called Monserrat manuscript of the *Historia general* at a monastery in Seville until such time as it could be published, the work consisted of a total of fifty books. These were divided not between the three volumes Oviedo had mentioned in 1535 but a staggering four.[9] The continued discoveries and conquests on the continent prompted Oviedo to accommodate this new information. So too, sections of the work devoted to New World nature underwent modifications, as they were expanded, corrected, and reorganized. Unfortunately, these efforts would not be fully appreciated until the mid-nineteenth century when the Spanish historian José Amador de los Ríos edited the three-part, fifty-book manuscript (1851–1855) under the auspices of the Real Academia de Historia.[10] For more than three centuries, access to Oviedo's magnum opus was limited to the 1535 edition.

The *Historia* (1535) marks a midpoint in Oviedo's writings on the Indies, coming roughly ten years after the publication of the *Sumario* and fourteen before his final edits to the Monserrat manuscript. As such, the work recalls its predecessor's treatment of the natural world at the same time as it anticipates subsequent articulations of this topic. Perhaps the most notable difference between the two printed works is that the *Historia* (1535) steps away from the broad coordinates of the *Sumario*, which surveys the flora and fauna of the Caribbean at large, to focus exclusively on the human and natural histories of the islands. Oviedo dedicates ten books (6 to 15) to the plant and animal life on Hispaniola, one each to San Juan (Puerto Rico), Cuba, Jamaica, and Cubagua (off the coast of Venezuela), respectively, and one chapter to Margarita (in the book on Cubagua). The disproportionate attention Oviedo devotes to Hispaniola reflects his personal circumstances at the time—by 1535 he had been a resident of Santo Domingo for over a decade. If the *Sumario* is rooted in Oviedo's experiences in Castilla del Oro, the *Historia* (1535) reflects his new life on Hispaniola and the knowledge of its flora and fauna he acquired as a result.

His personal ties to Hispaniola provide a convenient solution to the lack of biodiversity Oviedo observed among the islands.[11] Recall that similar redundancies in the *Sumario* prompt Oviedo to address species common to the islands and mainland in his account on Castilla del Oro, where they are present in greater numbers. He takes a comparable approach in the *Historia* (1535), in this case employing Hispaniola as a catchall for flora and fauna ubiquitous throughout the island chain. He justifies his brief account of nature on San Juan, explaining that the islands are alike "in many things . . . in the birds and the animals and fish and other particularities."[12] A similar logic informs Oviedo's decision to direct readers to the books on Hispaniola for further information on Cuba's wildlife.[13] He makes exceptions to this tendency for phenomena unique to particular islands, which he addresses as components of their geographies: the *palo santo* on San Juan, which Oviedo touts as a cure for syphilis, an oil well on Cuba, and a form of fishing particular to Jamaica.[14] For the most part, however, Hispaniola functions as the narrative fulcrum of the *Historia*'s natural history. This editorial strategy serves several purposes, allowing Oviedo to capitalize on his firsthand experiences and to avoid repeating the same information relative to each of the islands.

Despite the continuities between the *Historia* (1535) and its predecessor, the more restricted geography of the former presents Oviedo with a new set of challenges. Foremost among these is the absence of island fauna. The majority of the large mammals on *tierra firme* that Oviedo had described in the *Sumario* simply did not exist on Hispaniola or any of the neighboring islands.[15] What is more, the small animals that at one time had inhabited the island in large numbers—the guinea pig, hutia, and rat—had been hunted to near extinction by the first conquistadors, making it difficult for Oviedo to track any down.[16] These extenuating circumstances compel Oviedo to deviate from his standard, experiential approach to the natural world—a method that will be addressed in more detail shortly—and rely on informants' untrustworthy and spotty memories. In describing the *quemi*, Oviedo confesses that he has never seen the animal, but that "many allege that it has four legs and is as big as a medium-sized hound dog, and is the same brown color as the hutia and of similar proportions except the *quemi* is much bigger."[17] He depends on others to describe its taste, too, but he does not give their testimony much credence; given the hardships the first colonizers faced, Oviedo argues that "anything that they ate would have seemed delicious to them, even if it wasn't."[18] Similar difficulties beset Oviedo when attempting to describe the hutia, the *mohuy* (a kind of rodent), and a small dog he refers to as *perro gozque*. Without any experience of his own by which to evaluate these secondhand accounts, Oviedo shares what little information he gathers hesitantly.

These circumstances account for the brevity of *Book 12: On Animals* as well as its placement in the *Historia* (1535) relative to island flora. One of the shortest books in the work, its eight chapters are laid out over a mere two and a half folios. Whereas Oviedo had organized the *Sumario* to loosely follow his Roman predecessor Pliny the Elder's model, which begins with four-legged creatures and concludes with trees and medicinal plants, in the *Historia* (1535) he inverts this order entirely.[19] Animals no longer feature so prominently in the work, being represented in a single book placed after those devoted to flora (7 to 11) (and preceding books 13, 14, and 15, on animals of the water, birds, and insects, respectively).[20] Jesús Carrillo Castillo attributes this new arrangement to the prestige of Alonso de Herrera's *Libro de agricultura* (*Treatise on Agriculture*) (1513), which concludes with an account of domestic animals.[21] However, it is not until Oviedo began the process of revising and expanding the *Historia*, sometime after 1535, that he references Herrera's work.[22] It is much more likely, then, that Oviedo made the decision to reorder these books due to his need to rely on others for information about the animals on Hispaniola. In so doing, Oviedo chooses to foreground those phenomena about which he can speak most accurately and authoritatively—namely, the island's plant life.

The organization of the *Historia*'s natural history responds to a variety of considerations, as it had in the *Sumario*. Above all, Oviedo privileges the knowledge born of his firsthand experiences living on the island of Hispaniola, viewing other eyewitness testimony with suspicion. Despite the temporal and geographical distances separating the *Historia* (1535) from its predecessor, together they represent Oviedo's initial solution to the challenges of documenting this new reality.

The Hesperides and Classical Conceptions of the Natural World

The *Historia* (1535) elaborates on the *Sumario* not only in the scope of its natural history—albeit on a region, the Caribbean, largely unexplored in the latter—but also in the historical circumstances that form the backdrop to this narrative. Fundamentally, Oviedo's view that the Indies had been known in the distant past and ruled over by Hespero, the twelfth king of Spain, serves to legitimate Castile's political authority over these territories.[23] Nevertheless, Oviedo's attempts to demonstrate the persistence of these lands in classical and medieval sources also point to the basic premise underlying his conception of the forces molding the natural world at large. The notion of the inherent oneness of the universe, despite its embrace of a great diversity of lands and peoples, is the foundation of Oviedo's endeavors to give meaning to his experiences in the New World.

In *Book 2* of the *Historia* (1535), Oviedo celebrates Christopher Columbus as the "founder" of Spain's enterprise in the New World, a task the Genovese undertook for the benefit of "all those living and our successors."[24] Somewhat counterintuitively, he argues that the explorer ushered in a new era for Castile by laying claim to territories that it already possessed. Indeed, Oviedo alleges that the Indies had been found not once but twice before Columbus's inaugural voyage, albeit memory of these historic episodes had long been forgotten.[25] He attributes the Genovese's success in recovering these lands to his diligent study of ancient writings, which contained obscure references to these prior discoveries, as well as to the aid of divine providence.[26]

Turning to these classical authorities, Oviedo takes great pains to demonstrate the long-standing presence of the Indies in the ancient world.[27] He begins by reconstructing the circumstances of their penultimate discovery by a group of Carthaginian merchants, a tale that he ascribes indirectly to Aristotle through his reading of the Dominican Theophilus de Ferrariis (d. ca. 1493). According to Oviedo, the merchants ventured beyond the Strait of Gibraltar and into the Atlantic Ocean, where they encountered a deserted island inhabited only by wild beasts. So fertile was the soil and abundant its produce that the merchants decided to settle there. However, the Carthaginian government, fearing the great power other nations would acquire were they to discover and conquer the lands for themselves, ordered that travel to the island cease and those living there be killed.[28]

Oviedo hypothesizes that the land the Carthaginians stumbled across could very well have been Hispaniola, Cuba, or even part of the mainland, though by the end of the *Historia* (1535) he rejects the latter as a viable option.[29] His theory rests both on the logic of geography as well as the merchants' favorable impressions of the island, which coincide with his own high regard for Hispaniola and its wealth of natural resources.[30] Effectively, this tale allows Oviedo to complement his direct knowledge of the Caribbean with the weight of classical authority. Creating continuity between the past and the present, Oviedo imparts credibility to his assertions concerning the economic possibilities of these lands by setting his observations against the backdrop of this earlier encounter.

It is not the Carthaginians to whom Oviedo credits the initial discovery of the Indies, however, but Spain. Employing such classical and medieval authorities as the Roman geographer Statius Sebosus; Pliny the Elder (23–79 CE); Gaius Julius Solinus (ca. 250); the archbishop Rodrigo Jiménez de Rada (ca. 1170–1247); and Pseudo-Beroso, Annius de Viterbo (1432–1502), Oviedo seeks to legitimate the Crown's presence in the Americas by arguing that Spain ruled over these lands long ago. His first task is to demonstrate that the ancients knew the territories in question by a different name, the Hesperides. To support his claim that the Hesperides had once belonged to Spain, Oviedo traces their discovery to the reign of Hespero, the twelfth king of Spain, more than three thousand years ago.[31] He attributes the linguistic similarities between "Hesperides" and "Hespero" to the custom that arose in the wake of the Tower of Babel of naming kingdoms and territories after their rulers; therefore, Oviedo concludes, these lands owe their name to King Hespero. Columbus's arrival in the Caribbean restored Spain's centuries-old possession of these territories.[32]

Oviedo's professions of an ancient post-Diluvian Spanish Empire ostensibly provide a historical basis for the Crown's most recent territorial claims to the Americas.[33] They also reveal the ideological premise on which his understanding of nature at large rests.[34] Oviedo takes exception to Pietro Martire d'Anghiera, who dubs Spain's overseas territories a "New World." In point of fact, Oviedo retorts, the Indies are neither younger nor older than Asia, Africa, and Europe.[35] If classical authorities had provided the islands' geographical coordinates instead of describing their location relative to other land masses, he adds, news of their discovery would not have been lost and such egregious misnomers would have been avoided.[36] His contention that these lands are just as old as their Continental counterparts and constitute "part or half of the world" speaks to his belief in the inherent unity of the globe.[37]

Having existed since the beginning of the world, Oviedo is convinced that the Indies must have participated in the pivotal events of the history of Christianity, including the apostles' evangelization efforts. Oviedo quotes a statement by St. Gregory the Great in his *Commentary on Job* (578–595 CE) as confirmation that New World natives have received the word of God: "The Church has already preached in all parts of the world the mysteries of our redemption."[38] That indigenous communities do not conserve any memory of such an event does not dishearten him, ascribing this lacuna to forgetfulness.[39] For Oviedo, the very forms of nature in the Indies incarnate the Holy Spirit, serving as a lasting reminder of what he believes is the one true religion. One of the most notable "signs" Oviedo discovers are the crucifix-shaped leaves of the *higüero* tree, which stand as "testimony of the Cross that these people [the natives] have not been able to ignore."[40] The botanical realities of the Indies embody their divine creation and are material testaments of the Christian faith.

At the same time as Oviedo demonstrates that the Indies are part of a universal Christian narrative, he highlights their biological uniqueness. For as many similarities as there are between the species Pliny the Elder describes in the *Historia naturalis* and those that Oviedo observes in the Americas, the Spanish historian also points to phenomena about which his predecessor was ignorant or only partially informed. Yet, as we saw in Chapter 2, these discrepancies do not cause Oviedo to cast suspicion or doubt on Pliny's natural history. There would have been little need for him to do so since authorities from Pliny to Albertus Magnus (1200–1280 CE) subscribed to cosmological paradigms that accommodated the difference and novelty inherent in nature. One of these, the theory of the five zones, asserted a five-part division to the world based on the amount of heat received from the sun, which in turn determined the habitability of any given region. The cold zones beyond the Arctic and Antarctic Circles were deemed uninhabitable since they did not receive enough heat, whereas the torrid zone, located between the Tropics of Cancer and Capricorn, was too hot.[41] In these tropical regions, only geographical accidents, such as large bodies of water, were considered capable of creating the humidity necessary to generate life.[42]

Adopting this model, Oviedo attributes the habitability of the torrid Indies to Providence, which transformed these naturally hot lands into temperate ones thanks to the frequent rains as well as the abundance of natural sources of running water. Although Oviedo acknowledges that the ancients wrongly claimed all lands in the torrid zone to be uninhabitable, he does not challenge the theory giving rise to these assumptions. Instead, he points to a lack of knowledge that prevented ancient authorities from identifying the specific geographical features of the Indies that make them deviate from the norm.[43]

Oviedo's comments about New World plants and animals also reflect long-standing views on the importance of place in the propagation of life. In the Renaissance, Albertus Magnus's Christian theology was the primary vehicle that shaped humanist understanding of the influence of heavenly forces on natural matter. In *De natura loci* (*On the Nature of Places*), Albertus argues that celestial radiation imparts unique properties to every place on earth, resulting in innumerable natural spaces distinct from one another. Although places closer to one another tend to resemble each other more than those that are geographically distant, any given territory is composed of particular qualities that bestow on its inhabitants—humans, plants, animals, and minerals—distinctive traits.[44] In the *Sumario*, Oviedo articulates a similar understanding of the factors that endow New World nature with its singular features: "It is true that, according to the marvels of the world, creatures have different characteristics according to the different provinces and constellations where they are nurtured. While here plants are harmful and poisonous, there they are inoffensive and useful. Birds that in one province taste good, in another are not even eaten."[45] Oviedo does not reference *De natura loci* in his writings on the Indies, though he does attribute another work, which he refers to as *De propietatibus rerum* (*On the Properties of Things*), to the Dominican theologian. Instead, his understanding of the effect of the stars on the variety and novelty in the natural world likely speaks to the deep-seated influence of this philosophy on Renaissance thought. In the *Historia* (1535), Oviedo applies the same logic to explain differences in the appearance, habits, and function of certain phenomena that he observes and which Pliny also describes, from trees that do not lose their leaves to the formation and growth of pearls.[46] The biological parallels and discontinuities that Oviedo traces between species in the Indies and those described by Pliny reflect his views about the universal laws of nature and how they operate in the New World.

That the natural world is both one and many is a fact that Oviedo takes for granted. Far from questioning the paradigms of his predecessors, he functions within them.[47] Oviedo establishes the unity of the world as predicated on the twofold affirmation of the Indies as historically Castilian and inherently Christian: by endowing these lands with an ancient political lineage and positioning them within a religious narrative of universal creation, he concludes that this other "half of the world" has existed since the beginning of time.[48] Yet, within this unity, there is diversity: the peculiar character of these territories is attributed to their celestial and geographical bearings, from which emerge a natural world that is prodigious, varied, and unique.

Constructing the Natural World of the Indies in the *Historia* (1535)

The exceptional qualities of Spain's overseas colonies prompt Oviedo to make a case for the focused, deliberate study of their natural life. In the prologue to the *Historia* (1535), he lauds the abundance and diversity of the New World in an encomium that encompasses the full range of its offerings, from the mountains "more admirable and terrifying than Etna" to fruits, trees, rivers, and beaches.[49] Oviedo is as overwhelmed by the artistry and immensity of these lands as he is convinced of their economic potential. He champions a published account of the Indies as accomplishing a dual purpose: to disseminate information about their resources for the benefit of all and to celebrate Spain's mighty empire. News of these wonders, Oviedo adds, will allow other nations to learn about "these kingdoms and provinces, with their strange inhabitants and their customs, ceremonies and idolatries, as well as their abundance of diverse phenomena, [which] have remained detached from all that has been written from the beginning of time until the present."[50] Oviedo's conviction of the centrality of natural history to accounts of the New World is more than just a rhetorical flourish: he dedicates the majority of the books in the *Historia* (1535)—fifteen out of a total of twenty—to such topics, and in particular to the plants and animals of the Indies. More than any human actor, it is the vast wilderness that is the work's true protagonist. For Oviedo, the entire success of Spain's colonial endeavors rests on acquiring knowledge of this new reality.

In the *Historia* (1535), Oviedo wrestles with some of the fundamental questions that would continue to engage natural historians throughout the sixteenth century and beyond regarding the making of knowledge and experience. Foremost to these efforts is Oviedo's use of illustration to complement and enhance the written narrative. Although Oviedo intended image and word to be "read" in tandem, as visual objects woodcuts differ intrinsically from the text they accompany. For this reason, they merit attention in and of themselves, and will be considered in the following chapters. Here, we turn our attention to Oviedo's written account of New World nature in the *Historia* (1535) to examine the literary practices and strategies that characterize his approach to this topic.

Much as he did in the *Sumario*, Oviedo establishes his authority in the *Historia* (1535) as deriving from his personal experiences. He clearly delineates the limits of this knowledge for readers, distinguishing between his primarily descriptive task as natural historian and what he considers the more dangerous and useless conjecturing of natural philosophers. He recalls an anecdote—fabricated, though not acknowledged as such by Oviedo—from Gallensis Joannes's *Summa joannis valensis de regimine vite humane* (*The Complete Thoughts of John Valensis concerning the Regimen of Human Life*) (1496), detailing the circumstances of Aristotle's death. Frustrated that he could not understand the cause of the ebb and flow of the sea, the Greek philosopher drowned himself, crying out to the water, "Understand me, since I cannot understand you."[51] For Oviedo, the moral of this episode is that man should be content with what he is able to understand by the grace of God and not rely solely on his intellect.[52]

His words of caution echo the warnings issued by Saint Augustine in his *Confessions* (397–400 CE) concerning the dangers of curiosity "purely for the sake of testing boundaries—not because it wants to endure hazards, but because of its craving for experience and understanding."[53] Augustine considers this desire for knowledge for knowledge's sake a most pernicious temptation because it "sets in motion investigation into the secrets of nature (which are beyond our ken)."[54] Similarly, Oviedo distances himself repeatedly from the model of the natural philosopher, insisting that he will not consider "the effects from which proceed the novelties that I recount, because I am neither such a philosopher as to be able to understand them nor do I wish to be detained in arguments. Rather, I will say what I have been able to understand or what I have felt in these materials, according to what is visible."[55] Oviedo is convinced that idle musings into natural causes will most certainly end in failure; to probe beyond the limits God has placed on human knowledge is not an invitation to the curious but rather a futile and ultimately dangerous venture.

Experience, in the sense it is employed in the *Historia* (1535), is diametrically opposed to the passive, internal search for meaning that characterizes the discipline of natural philosophy. Instead, Oviedo conceives of this practice as founded on active, engaged contact with the external world. As per his view, such interactions generate knowledge through prolonged and repeated exposure to their focus of study. Oviedo's sustained examination of New World flora and fauna differentiates his gaze from that of other eyewitnesses, whom Oviedo criticizes for making claims in haste and without sufficient evidence.

These tensions are most evident with regard to the tree on Hispaniola that popular opinion has dubbed the terebinth, after its namesake in Pliny's *Historia naturalis*. After a thorough examination of the tree, however, Oviedo questions the resemblances between the two, observing that "these [New World trees] appear very different from those that [Pliny] describes."[56] Although some specimens share features in common with the terebinth, he points out that for every such similarity there are numerous differences. For Oviedo, the confusion over the identity of the so-called terebinth tree is the fault of those who "seeing only one indication of likeness, and having no experience in such things, give [the tree] the name as if it had all the characteristics that Pliny lays forth."[57] In contrast to these quick, superficial verdicts, Oviedo's misgivings come from having observed the tree repeatedly, in a variety of contexts: "Many times [I] inquired as to this tree, finding [myself] on these roads and in these forests and in different places on these islands and *tierra firme*."[58] He does not completely dismiss the notion that the tree could be terebinth—after all, Pliny highlights the great diversity within the species. Nevertheless, Oviedo cautions against using this name for the plant, believing the truth will emerge only through further exploration.[59]

The case of the pseudoterebinth illustrates how, for Oviedo, to truly see—that is, to use one's eyes to generate accurate information—is to undertake a series of encounters with the same object of inquiry, over time and in different spaces. This approach to the natural world is not new, forming part of a centuries-old Aristotelian tradition in which a basic understanding of any given phenomenon is reached through repeated sensory experiences.[60] It is, however, the first time that New World species are subjected to such systematic examination.[61] Although this method does not produce immediate results, Oviedo defends its conclusions as being more well-informed than those of observers who pass judgment based on knowledge that is insufficient or secondhand.[62]

The implications of this approach for an understanding of the natural world of the Indies are visible in many of the entries on plants and animals in the *Historia* (1535). Since Oviedo envisions the creation of knowledge as a multistep process, consisting of the synthesis of discrete moments in time and space, it follows that his vision of New World nature subsumes the particular to the general. His accounts of yuca (cassava) and corn are two of the many that illustrate this rhetorical strategy. Oviedo begins by qualifying the yuca's height and girth as "somewhat taller than a man but others which are much smaller, as thick as two fingers, and some are thicker and others thinner."[63] He employs similar language regarding the optimal harvest time for corn, stating that generally it is four months, though in some places it is as early as two or three months, and in Nicaragua, as soon as forty days after planting.[64] In both instances, Oviedo identifies typical patterns of growth, which he then qualifies by acknowledging divergence from this norm. The result of such descriptions is a composite vision, one that strives to identify standard behaviors while at the same time admitting difference and variation.

Occasionally, Oviedo recounts anecdotes detailing his encounters with a particular specimen. When he does so, he is exacting in their description, often including specific measurements to support his claims. The snake that he happened upon on the island of Hispaniola in 1515 measured "more than twenty feet long and the widest part of it was bigger than a closed fist."[65] When, in Panama in 1529, a frigate bird (*rabiohorcado*) met with an untimely death after attempting to eat some sardines drying in the sun, it underwent a similar examination. Oviedo recalls that, "many men . . . extended their arms trying to see if their arm span reached the tips of the bird's open wings, but they were more than four fingers away."[66] Eventually, the bird's feathers were plucked and Oviedo held the creature in his hands, surprised at how small its body was—barely bigger than a dove's—relative to its enormous wingspan. He contends, "I would not have been able to believe what I saw through my own experience if anyone had told me about it."[67] The study of these specimens provides data that Oviedo uses to formulate conclusions regarding the species at large. At the same time, each of these experiences is unique, as Oviedo recognizes these creatures as the products of their geography and circumstances.

Oviedo's emphasis on the uniformity and regularity of the natural world distinguishes his writings from those of his Baroque successors, many of whom placed value instead on nature's abnormalities. Authors and fellow residents of the New World, such as Antonio de León Pinelo (1590–1660) and the Jesuit priest Alonso de Ovalle (1603–1651), maintain that it is the marvels and monstrosities that most reveal God's omnipotence.[68] Oftentimes, such aberrations were understood as defying the laws of nature, prompting speculation as to their allegorical or symbolic meaning as divine signs.[69] In contrast, this discourse on singular wonders is largely absent from Oviedo's *Historia* (1535), where God's artistry is apparent in the order and arrangement of the visible universe. Oviedo insists that "in nature there is nothing wrong nor badly composed, because its maker could not err nor make anything unsuitable nor without benefit."[70]

Even the most savage plants Oviedo encountered—what we know today collectively as cacti—embody this systematic order. Novelties for Oviedo and his readers alike, their appearance both fascinates and repels him; he describes one particular species of tall cactus as so "spiny and savage that there is no part of the plant that can be touched [that does not have] these very fierce spines."[71] Despite its hostile, unwelcoming physical form, Oviedo observes that "nature has arranged them [the spines] well and at a certain distance from one another with much order and measure in their composition."[72] As much as Oviedo puzzles over these strange beings, they are not anomalies: like all other plants, they reflect divine creation, their spines arranged just so. Oviedo is not interested in nature's irregularities but rather what is most commonplace and frequent.

Hierarchies of Knowledge: Native Knowledge and Informants in the *Historia* (1535)

In the face of the many challenges presented by these new species, Oviedo lauds his own efforts to procure information about them firsthand. His eyewitness observations serve as a retort—at times implied and at other times explicit—to Pietro Martire d'Anghiera's popular *De orbe novo decades* (*Decades of the New World*), the complete edition of which was published five years before the *Historia* (1535), in 1530. Having never set foot in the Indies himself, the Italian humanist relied on returning travelers for information. Conversely, as we have seen, the knowledge Oviedo acquires through his hands-on interactions with New World flora and fauna constitutes the backbone of the *Sumario* and the *Historia* (1535).

Yet Oviedo is but one individual, whose familiarity with the natural world is necessarily circumscribed to his limited reality. An agriculturalist, Oviedo has an intimate understanding of the planting, harvesting, and culinary preparation of certain crops; he explains that it is thanks to the twenty years he has spent growing corn for his household that he knows so much about cornbread.[73] But, in general, Oviedo is often faced with situations in which he is not able to provide a comprehensive account of a given species on the basis of his experiences alone. In these instances, Oviedo turns to other sources to supplement his observations. His informants—ranging from indigenous inhabitants to sailors, doctors, and pharmacists—provide him with insight into the properties and uses of a wide array of plants and animals. Answers to Oviedo's queries are not always forthcoming, however. The voiced omissions and silences in these instances speak to native resistance to Spanish colonization as well as sacred spheres of knowledge. Although scholarly discussions of Oviedo's writing tend to privilege his face-to-face encounters with the natural world, these local information networks, too, are crucial to his work. To render intelligible what he sees and experiences, Oviedo must complement his firsthand impressions of New World flora and fauna with the expertise of others.

Unsurprisingly, Oviedo owes much of his familiarity with life overseas to his interactions with indigenous communities, a position that mirrors the situation at large in the colonies.[74] Native lore permeates the *Historia* (1535): Oviedo acknowledges having observed and inquired into particular indigenous practices of preparing foodstuffs, such as dishes made with corn and yuca on the islands and *tierra firme*.[75] He also reports on indigenous medicinal remedies, describing how the natives of Cueva remove scars with the fat of an unnamed bird and detailing a technique for purging that the Spanish "learned from the Indians."[76] Oviedo admits to having acquired other practical knowledge from indigenous peoples too, such as the fact that the roots of *hobo* trees contain water and that *cocuyos*, or fireflies, serve as living lanterns.[77] Since Oviedo does not always identify the sources of the information he includes in the *Historia* (1535), it is likely that the work is indebted to indigenous knowledge in ways that he does not recognize or perhaps is not aware of. Nevertheless, its presence is felt throughout the *Historia* (1535), indicating the degree to which Spanish settlers in the New World depended on native informants to successfully adjust to life there.

The reliance of colonists on indigenous authorities is nowhere more strongly felt than in Oviedo's account of agriculture in the Caribbean. Oviedo understands the natives to have been the first to domesticate this wilderness, and therefore he attributes to them much useful information on the subject. He describes the practice of sowing corn as one the Spanish inherited from "the natives, and they do it in this way," and he praises the cotton crops that they once cultivated on Hispaniola.[78] Oviedo views these efforts to tame and transform nature's offerings as a defining feature of humanity at large. Although he comments approvingly on the taste of wild grapes on Hispaniola and in many parts of *tierra firme*, he adds: "so I believe that all wild grapes were like this, in the beginning, from which originated those that were cultivated and improved."[79] Unassisted, nature might produce fruits fit for human consumption, but Oviedo argues that it is only when domesticated that they reach their full potential. The exception to this rule appears to be medicinal plants and herbs; Oviedo offers no instructions for growing them, indicating that he likely subscribed to the widespread belief in the Renaissance that remedies were most effective when harvested from the wild.[80] For the most part, however, Oviedo maintains that it is thanks to man's labor that plants truly thrive.

While Oviedo does not discount the benefits of indigenous agricultural practices, nevertheless he maintains that they need refinement. Specifically, he looks to the European farming tradition to improve on these homegrown methods. He explains the kinds of interventions he has in mind: "Where nature produces some of these things, they will do better for help of man in irrigating and in other such tasks of which men learn the secrets of agriculture, such as grafting, pruning, [using] manure and irrigating, each in its time."[81] Techniques such as these, enumerated in agronomic handbooks by classical and contemporary authors alike, are virtually synonymous with Old World agricultural practices. While Oviedo credits the natives with having taught the Spanish how to sow seeds, he adds, "The Christians do it much better on account of their plowing the land where possible."[82] Settlers have also begun to cultivate previously wild plants: Oviedo points to the domesticated *cañafístola* and *hicaco* trees, which yield more abundant harvests than those growing unaided in the fields and forests.[83] He mentions too the expert opinion of certain agriculturalists, who maintain that walnut trees would produce better fruit if grafted.[84] For Oviedo, indigenous husbandry does not maximize the potential of the land as much as the farming methods colonists have brought with them from overseas.

Oviedo understands the limitations of native agronomy, as well as other types of knowledge related to nature, as residing in these communities' lack of a written history. He identifies the *areítos*, which he defines as song-and-dance ceremonies, as the principal means by which the Taínos and other indigenous societies remember the past. Oviedo describes the role of the *areíto* in recalling major events of political and civic life, including moments of war and peace, the genealogies and accomplishments of leaders, and weather-related occurrences.[85] He does not, however, ascribe indigenous knowledge of the natural world to this oral tradition. Instead, he locates this wisdom primarily in natives' firsthand experiences of the healing properties of plants: "They know many herbs and they have experimented with them for this and other illnesses."[86] It is in this kind of intelligence born by observation where Oviedo finds the answer to how the natives became acquainted with the bow and arrow. Although Pliny credits this invention to a sole author, the son of Jupiter, Oviedo hypothesizes a much earlier origin in the biblical story of Lamech, the father of Noah. As such, he considers the bow and arrow to be the oldest human weapon and the most instinctual defense, concluding, "Naturally these savage people were able to discover them."[87] By constructing an argument in which knowledge of the bow and arrow is so innate as to be naturally occurring, Oviedo can speculate that they learned of this weapon from interacting directly with their environment.

While Oviedo upholds the importance of such hands-on encounters with nature for generating knowledge—after all, this is his modus operandi in the *Historia* (1535)—he also points to the dangers of arriving at faulty conclusions on the basis of observation alone, without considering the laws of nature. In Oviedo's view, indigenous knowledge is especially prone to these kinds of errors, which occur when natives "misread" nature's operations. This is the case with the *peje reverso*, a fish used to hunt tortoises and manatees due to its ability to affix itself to its prey. Oviedo describes how the natives talk to the *peje reverso* as they release it into the water, believing that only through this act will the fish accomplish its mission successfully. He counters this explanation by sharing what he has observed in the ocean, where he has seen the fish behave in the same fashion independently of human intervention. Dismissing as foolishness the notion that the *peje reverso* heeds human commands, Oviedo associates the natives' ignorance with their "not understanding that this is a property of nature."[88] A fish can no sooner comprehend human language as man can communicate with fish; the *peje reverso* obeys its instincts, not the wishes of its human handlers.

Oviedo maintains that it is necessary to confirm knowledge generated by the senses via written tradition to avoid making such mistakes. Indeed, he envisions his own writings as contributing to these efforts, stating in the

Sumario: "What most conserves and sustains the works of nature in man's memory are the histories and books in which they are written, and those that are taken to be more truthful and authentic are those that are written by he who, with prudent understanding, has traveled the world and seen with his eyes."[89] Eyewitness observation is a necessary but preliminary step toward documenting the natural world. The books immortalizing these wonders contribute to a system of shared knowledge about them. Over time, the information they contain is augmented, improved, and corrected by future generations of writers. Oviedo's readings of Pliny's *Historia naturalis* epitomize this cycle, as he complements the Roman historian's work with his knowledge of New World nature. So too, Oviedo envisions his own work as part of an ongoing, collective effort to document the natural and human histories of the Indies.[90] For Oviedo, the creation of knowledge necessitates a system of writing that allows words to bridge vast reaches of time and space; native knowledge of New World nature remains excluded from these processes.

The oral character of indigenous knowledge presents numerous difficulties for Oviedo, the most pressing being that native informants are not always willing to answer his questions. Lamenting the challenges he faces, Oviedo writes, "Since the native elders are dead, they have taken with them their knowledge of these properties and secrets of nature, or rather, those that were experimented and known by them. And all that can be said now is little and not well understood because this [present] generation is so miserly of the little they know, and for no reason or enticement will they reveal these things, especially that Christians could use, which are medicinal."[91] The demise of native elders and the unwillingness of those living to divulge information represent insurmountable obstacles for Oviedo.[92] It is possible that Oviedo alludes in this passage to the population decline on the islands occasioned by the presence of the Spanish, which likely contributed to the elders' deaths and the loss of traditional knowledge. He does not, however, query the relationship between natives' knowledge of flora and fauna and their spiritual beliefs. Nevertheless, it is likely that the curative powers of plants were tied to ritual practices, and perhaps even to certain shamanic leaders for whom this kind of sacred knowledge was often reserved, making native informants averse to revealing such secrets to Oviedo.[93] Whatever the case may be, later natural historians also encountered similar indigenous reticence in their attempts to document the healing properties of plants, among them the physicians Nicolás Monardes (1493–1588) and Francisco Hernández (ca. 1514–1587).[94]

The natives' reluctance or inability to communicate what they know of New World nature is one of the reasons that Oviedo strives to document these practices and traditions in the *Historia* (1535). Assuming most of his readers would never have seen an iguana, let alone eaten its flesh, he includes instructions for how to fry its eggs. Somewhat counterintuitively, neither butter nor oil will have the desired result; rather, the egg must be cooked in water. Oviedo's reason for including this information is simple: "I want to advise anyone residing in these parts who reads this (if the natives are lacking like they are now) of the manner and art that is necessary to fry iguana eggs."[95] A similar logic informs Oviedo's description of the two processes by which the natives on Hispaniola alternately boil and cool the liquid extracted from yuca to make sweet liquor and acidic vinegar, respectively. Oviedo bemoans the fact that "few Indians know how to do these experiences, because the elders are dead and because the Christians have no need of this vinegar when there are so many oranges and lemons on the island, and even less of sweet liquor with sugar so abundant. And so, it has been forgotten what the yuca and its juice were used for in these two cases."[96] Oviedo recognizes that the colonists' access to other products has made these preparations superfluous. Nevertheless, he shares what he knows of them with readers, fixing in written form this tradition for posterity's sake.

In addition to native testimony, Oviedo acquires information from other informants too, albeit to a lesser extent. His references to these sources indicate the wide range of local experts that Oviedo enlists to augment what he learns from his firsthand experiences. Agriculturalists give Oviedo tips on crop cultivation; sailors and fisherman, insight on fish and navigation; and herbalists, pharmacists, and medical doctors, the proper names for certain plants.[97] In the case of San Juan and Cuba, islands with which he is less acquainted, Oviedo relies on eyewitness reports from Juan Ponce de León, Diego Velázquez, and Pánfilo de Narváez, among others. The testimony from such a diverse array of explorers, artisans, and professionals provides much-needed supplementary material for Oviedo's descriptions of the flora and fauna of the Indies. By drawing on these different local networks, Oviedo puts into practice the Spanish humanist Juan Luis Vives's entreaty to scholars in his *De tratendis disciplinis* (*On Education*) (1531) not to be ashamed to "enter into shops and factories, and to ask questions from craftsmen, and to get to know the details of their work."[98] Oviedo views these interactions as vital to the comprehensive coverage he seeks to offer readers as a natural historian of the Indies.

The Legacy of the *Historia* (1535)

Although the *Historia* (1535) is not Oviedo's first writing on New World nature, it is here that he gives voice to the philosophical framework which informs his account of life in this other hemisphere. As old as time and as varied and diverse in their flora and fauna as any other place on earth, the Indies reflect a world that is both one and many. The *Historia* (1535) represents the first time Oviedo articulates his approach to the making of knowledge and experience. His exhaustive study of the New World plants and animals that he encounters is nothing less than a homage to God; even the most frightful species have their own internal logic that reveals the hands of their creator. To execute his task to the best of his abilities, Oviedo turns to a host of local informants to make understood different facets of this new reality. These features are the hallmarks of Oviedo's approach to writing the natural history of the Indies and the foundation of the *Historia general*'s enduring legacy.

CHAPTER 5

Illustrating Nature

The Visual Dimensions of Flora in the *Sumario* and the *Historia* (1535)

Of all the woodcuts in the *Sumario* (1526) and the *Historia* (1535), that of the pineapple has taken on a life of its own as the quintessential representation of New World nature (Figure 3). An autonomous species of the Americas, the image of this rotund and scaly fruit in the *Historia* (1535), balancing precariously on its thin stem, simultaneously celebrates the new knowledge resulting from Spain's conquest and colonization and demonstrates the challenges of communicating such wonders to readers an ocean away.[1] Interspersed with the praise Oviedo lavishes on the pineapple is frustration at what he deems his futile efforts to do justice to the fruit: "The tongue cannot make understood the reason and image of this fruit so that it satisfies the details required in this case. And because I do not know how to draw as I would like, so that in addition to words the vision of the reader could participate in this truth, I will put here [a drawing] as best I can in the margins. Since the drawing isn't colored and it is lacking in skill, I cannot fulfill the demands of the material to make it understood completely."[2] While the dilemma presented by graphic representation is particularly acute in this case—after all, the pineapple is Oviedo's favorite fruit—the struggle of giving visual form to New World nature is one he often faces. As much as Oviedo embraces illustration as an integral component of his natural history of the Indies, throughout his career he grapples with how to best employ this artistic medium. The difficulty he expresses in attempting to draw the pineapple is only one example of how Oviedo reflects on the uses and shortcomings of visual expression in the *Historia* (1535).

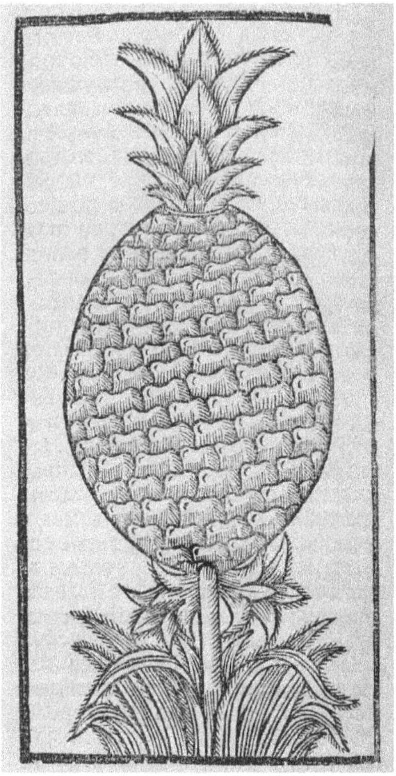

Figure 3. Pineapple. *Historia* (1535), lxxvi v [bk. 7, chap. 8]. Beinecke Rare Book and Manuscript Library, Yale University.

The wealth of illustrated material in the *Sumario* and the *Historia* (1535) is unprecedented in the early literature of colonial Latin America. The former includes four woodcuts, and the latter twenty-nine.[3] No surviving manuscript of either publication exists, making these woodcuts the earliest extant images in Oviedo's visual repertoire. Even though their creation cannot be directly attributed to him, there is no doubt that the artisan(s) who traced and carved them relied on sketches Oviedo supplied. The striking likeness of many of the drawings in the Monserrat manuscript (1535–1549) to these earlier woodcuts indicates that they were all made by the same hand. Furthermore, the marginal instructions Oviedo left regarding the handling of the illustrations in the Monserrat manuscript reveal the degree of his involvement in the printing process. One could imagine that Oviedo would have been equally insistent that the woodcuts in the *Sumario* and the *Historia* (1535) be executed according to his specifications, given his claims to have shouldered all printing costs.[4]

Woodcuts were common in the early sixteenth century, especially in popular reading material.[5] Yet publications on the New World prior to the 1530s did not follow this trend: relatively few, they did not often employ illustrations, and certainly not of the kind and quantity found in the *Sumario* and the *Historia* (1535). The earliest such account, and one of the most widely printed—the letter Christopher Columbus wrote to Luis de Santángel announcing the successes of his first voyage—did not contain any visual material in its first edition (Barcelona, 1493).[6] Although subsequent editions and translations of the letter into Latin (Basel, 1493/1494) and Italian verse (1493/1495) were illustrated, the scenes displayed more closely approximated fantasy than reality; loosely inspired by Columbus's account of his voyage, the artisans who created the woodcuts did not hesitate to recycle elements from preexisting images when needed.[7] Another early woodcut depicting a cannibalistic feast among the Tupinambá (Augsburg, 1505) similarly combined artistic license with a touch of realism based on models sent to Europe of the Brazilian natives' feathered attire.[8]

The woodcuts in the *Sumario* and *Historia* (1535) mark a radical departure from their predecessors in a variety of ways. Most importantly, they reproduce Oviedo's gaze, being based on sketches that he created of New World phenomena in situ. Then too, from the very outset Oviedo conceived of illustration as indispensable to his writings on the Indies; notwithstanding his constantly evolving approach to graphic representation, Oviedo remains convinced of the need to accompany text with image. His dependence on illustration is particularly salient with regard to the botanical realities of the New World: images of plants constitute roughly 50 percent of the woodcuts in both the *Sumario* and the *Historia* (1535), two in the case of the former and fifteen in the latter.[9] Not until the publication of Nicolás Monardes's *Historia medicinal de las cosas que se traen de nuestras Indias occidentales* (*Medicinal History of the Things that are Brought from our Occidental Indies that are Useful in Medicine*), in 1565, would other efforts be made to visually document flora of the Indies on a similar scale.

Yet, despite the singularity of these illustrations, they have not received the scholarly attention they deserve. In part, this is due to the editorial interventions of the Spanish historian José Amador de los Ríos (1816–1878) upon preparing the first complete edition of the *Historia general* (1851–1855). In Ríos's attempts to modernize the woodcuts, as well as the later, hand-drawn images from the Monserrat manuscript, he "distorted and bowdlerized Oviedo's originals."[10] So too, Ríos placed the illustrations in an appendix to the

Historia general, disrupting the close-knit relationship between text and image so vital to Oviedo's historiographic endeavors. The most popular twentieth-century edition of the *Historia general* (1959), edited by Juan Perez de Tudela Bueso, followed suit.[11] For well over a century, scholars have relied on these two editions, often unaware of how radically the editors altered Oviedo's illustrations and diminished their importance.[12]

Several of the first scholars to comment extensively on Oviedo's role as a natural historian, Enrique Álvarez López (1897–1961) and Antonello Gerbi (1904–1976), almost entirely disregard the woodcuts and drawings in the *Sumario*, *Historia* (1535), and the Monserrat manuscript. Gerbi mentions them only briefly and Álvarez López does not allude to them at all, likely because he does not consider them part of natural history proper.[13] Daymond Turner counters this trend: he advocates for a more robust scholarly engagement with Oviedo's visual oeuvre, referring to it as "one of the neglected treasure troves of the period of American discovery."[14]

Jesús Carrillo Castillo and Kathleen Ann Myers are among those who have heeded Turner's appeals.[15] Their respective treatment of Oviedo's visual corpus exemplifies the various possibilities for its study. Carrillo Castillo's sweeping survey of Oviedo's illustrations of flora and fauna spans four centuries and considers the editorial interventions of sixteenth-century publishers and translators of Oviedo's writings on the Indies into Italian, English, and French, as well as those of Juan Bautista Muñoz (1745–1795) and Ríos. In contrast, Myers is more selective, focusing almost exclusively on the illustration of the pineapple as epitomizing the difficulties Oviedo encountered in attempting to give visual form to New World novelties.

A different approach is needed, however, to adequately capture the depth and breadth of the woodcuts of plants in the *Sumario* and the *Historia* (1535), as well as the multifaceted interplay between text and image in the latter. In addition to Ríos's editorial decisions, other factors account for the lack of attention to the images in Oviedo's printed works in particular. Above all, the later hand-drawn images in the Monserrat manuscript overshadow their predecessors in scale and beauty, making them a greater focus of scholarly interest. Yet to overlook the woodcuts in the *Sumario* and the *Historia* (1535) is to neglect the very foundation of Oviedo's artistic program, the key to understanding his relationship with visual imagery during his first decades in the Indies and how it develops over the course of his career.

Also problematic are Oviedo's avowals of mimesis—that is to say, of having sketched certain plants and animals from life—which have often led scholars to focus on *what* is represented rather than *how* or *why* Oviedo chooses to represent it.[16] The tendency to ascribe a kind of objectivity to the visual material in the *Sumario* and the *Historia* (1535) is not unique to these works; those of other sixteenth-century illustrators of nature, also professing to have sought inspiration directly from their surroundings, have been similarly received. Increasingly, however, efforts have been made to challenge the seemingly transparent claims of naturalism, the guiding aesthetic principle of both early modern art and science. Carmen Niekrasz and Claudia Swan point to the fact that illustrations made *ad vivum* (from life) do not all look alike, despite being grounded in this common assertion.[17] This new generation of scholars urges a more discerning approach to early botanical prints, one that finds merit in the various techniques employed in their creation rather than the perceived accuracy of the phenomena portrayed.[18] Far from reflecting a neutral, impartial reality, these illustrations embody a wide range of perspectives and motivations.

Recently, several scholars of colonial Latin America have begun to conduct a similar reevaluation of the Spanish American visual archive. Most notably, Daniela Bleichmar employs this methodology to examine a corpus of twelve hundred nature illustrations produced during a series of scientific expeditions to the Americas that took place in the late eighteenth and early nineteenth centuries. She argues that each individual image embodies "multiple observations, decisions, negotiations, and types of expertise."[19] Like Niekrasz and Swan, Bleichmar challenges the seeming objectivity of these paintings, demonstrating how they reflect the presence of diverse actors involved in questions of representation.[20] Despite the vastly different circumstances in which Oviedo embarked on his historiographical project, the illustrations he created engage with similar issues. Oviedo, too, faced myriad decisions in determining what to depict visually, and how. His lifelong attempts to give graphic form to his experiences in the Indies make him a precursor to the artists who participated in these later scientific expeditions, as well as all other attempts to render this reality visible.

Representing Nature in Early Modern Europe

Oviedo's decision to illustrate New World flora and fauna speaks to larger shifts occurring in early modern scientific circles involving the study of nature. These changes took their cue from the classical impulse of the Italian

Renaissance toward artistic naturalism. As we saw in Chapter 2, just as artists such as Leonardo da Vinci aspired to imitate and even transcend nature's offerings, so too did humanists begin to consider their observations of the external world as necessary complements to Greek and Roman scientific knowledge. Although initially these scholarly efforts assumed written form, slowly those who undertook this pursuit began to embrace the possibilities of visual representation as well.

The works hailed as emblematic of this new trend were published contemporaneously with Oviedo's *Sumario* and *Historia* (1535): the first part of Otto Brunfels's *Herbarum vivae eicones* (*Living Plant Images*) came off the press in Strasbourg in 1530 and Leonhart Fuchs's *De historia stirpium commentarii insignes* (*Remarkable Commentaries on the History of Plants*) in Basel in 1542. Oviedo does not appear to have been acquainted with these men and their works, nor they his. Nevertheless, Oviedo shares with them an awareness of the ways in which the tools and techniques of artistic naturalism could be applied to the study of plants. These similarities are most apparent in the tight-knit relationship that Oviedo and Fuchs, independently of one another, establish between text and image in their writings, as well as the reasons they voice for including the latter. Despite the example set by their artistic peers and predecessors, the task of illustrating nature for the purpose of observation and inquiry presented unique challenges to this first generation of natural historians. The parallels between Oviedo's struggles to visually depict his experiences in the New World and those of these northern European humanists are as revealing as their differences, attesting to the circumstances in which each, in his own way, tackled this shared undertaking.

The works of Oviedo, Brunfels, and Fuchs form part of a long tradition of plant illustration in manuscript and print. Its earliest extant manuscript, the *Juliana Anicia Codex of Dioscorides* (also referred to as the *Codex Vindobonensis*) (512 CE), contains roughly four hundred pages of colored paintings of plants, rendered naturalistically.[21] Yet, in the same way as the practice of copying and recopying the writings of Greek and Roman authorities introduced hundreds and thousands of textual errors into the corpus, so too did similar distortions come to characterize efforts to reproduce these plant images (Figure 4). By the time the first herbal incunabula were published, in the 1480s, the naturalism of classical illustrations had vanished, leaving behind "crude, stylized, and derivative figures."[22] Ostensibly used as guides for identifying plants and their remedies, a typical entry in an herbal included a woodcut alongside a written description. Given the degraded nature of the

Figure 4. *Abrotanus. Hortus sanitatis* (Moguntaie: Jacobus Meydenback, 1491), 2r. Courtesy of the Linda Hall Library of Science, Engineering and Technology, Kansas City, Missouri. The woodcuts in this work were based on those of the *German Herbarius* (1485). See Gill Saunders, *Picturing Plants: An Analytical History of Botanical Illustration* (Berkeley: University of California Press, 1995), 18.

images, however, it is unlikely that they were much help in identifying the species they depicted. It is to be assumed that the herbalists, medical practitioners, and apothecaries who consulted these books knew enough about the herbs referenced to locate them unassisted in the wild. What does not leave any room for doubt is their popularity: herbal incunabula instantly became bestsellers of the burgeoning printing press. Two of the most widely read, the *Latin Herbarius* (1484) and the *German Herbarius* (1485), were published in multiple editions and translations. By 1500 these, along with other herbals that followed in their stead, had generated a total of thirty-two editions.[23] It is to be expected that Oviedo, Brunfels, and Fuchs were acquainted with this literary tradition and saw their own works as improving on those of their fifteenth-century predecessors.

Indeed, the new relationship that artists and humanists sought to forge directly with nature had discernible effects on the print tradition of plant illustration. While Oviedo was busy documenting New World flora and fauna, Brunfels explored German plants in the *Herbarum vivae eicones* (1530–1536). Its title, meaning "living plant images," points to the source of its novelty: the detailed woodcuts, each occupying a full folio (Figure 5). Created by the German artist Hans Weiditz, a pupil of Albrecht Dürer, they are the product of intense study of real-life models. Weiditz even went as far as to reproduce the idiosyncrasies—a bent stem, a ripped leaf—of the individual specimens he observed.[24] This innovative approach to plant illustration stands in stark opposition to Brunfels's commentary. In contrast to the firsthand observations that Weiditz employed to create his drawings, Brunfels's written account is little more than a compilation of information taken from classical and contemporary authorities. Brunfels also neglected to mention Weiditz's woodcuts anywhere in the work, resulting in an uncoordinated relationship between word and image.[25]

It was another German physician, Leonhart Fuchs, who managed to unify text and image into one coordinated program. In *De historia stirpium commentarii insignes* (1542), Fuchs employs illustrations deliberately to make his written account understood, just as Oviedo does in the *Sumario* and the *Historia* (1535). So comprehensive and thorough is Fuchs that the resulting illustrations are decidedly not lifelike (Figure 6). To depict as many aspects of a plant as possible, Fuchs showcases conflicting stages of growth simultaneously.[26] As we will see shortly, this method is diametrically opposed to the more selective glance that Oviedo trains on New World flora, underscoring the diverse visual strategies available in the early sixteenth century to portray plants from life.

Notwithstanding these stylistic differences, fundamentally Fuchs and Oviedo understood the function of plant illustration in much the same way. In an argument reminiscent of Leonardo da Vinci's opinions on painting, which were explored in Chapter 2, Fuchs maintains that an image conveys information more immediately and completely than the written word does. He asks, "Who in their right mind would condemn a picture which, it is clear, expresses things much more clearly than they can be described with any words of the most eloquent men? . . . It is certain that there are many plants

Figure 5. *Diapensia.* Hans Weiditz and Otto Brunfels, *Herbarum vivae eicones* (Strasbourg, 1530), 80v-81r.

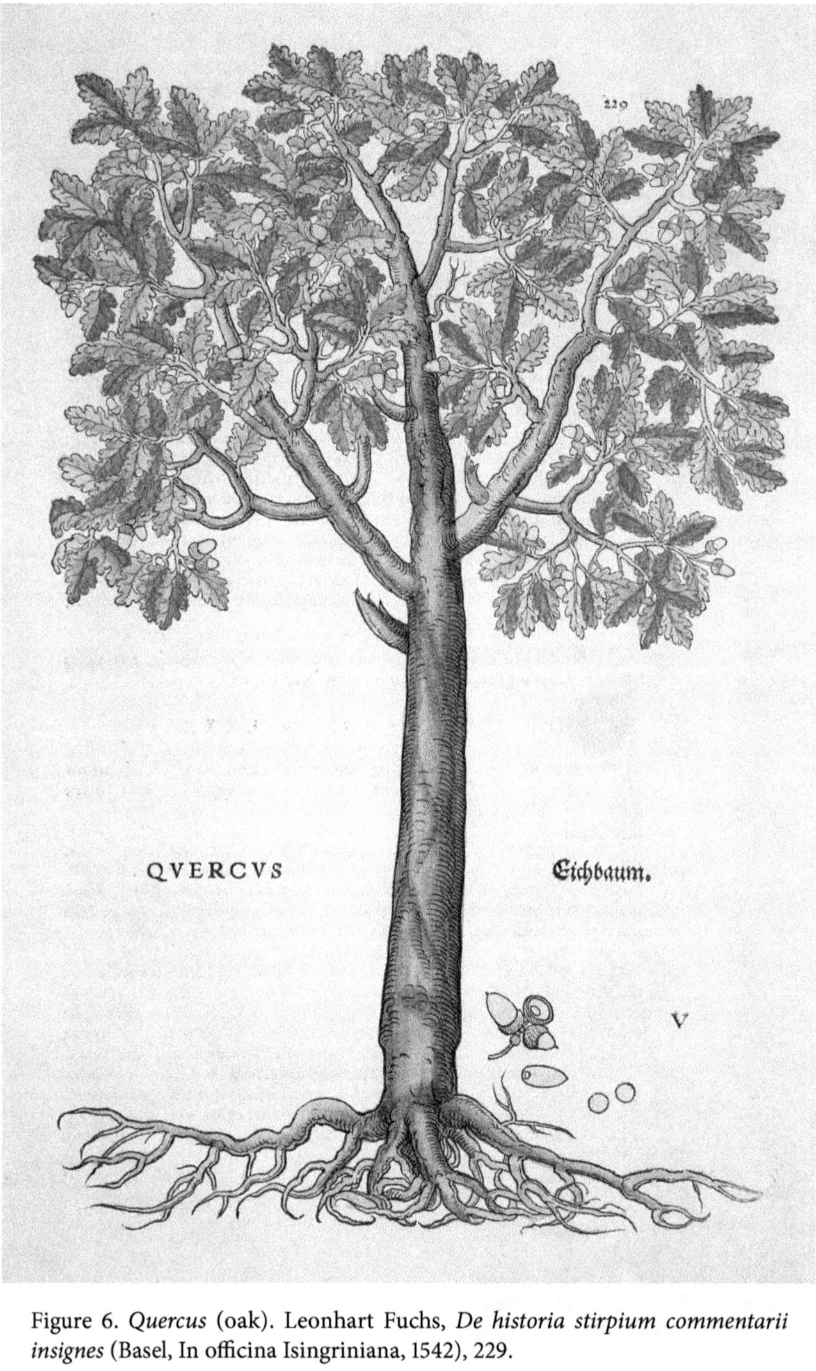

Figure 6. *Quercus* (oak). Leonhart Fuchs, *De historia stirpium commentarii insignes* (Basel, In officina Isingriniana, 1542), 229.

which cannot be described by any words so as to be recognized, but which, being placed before the eyes in a picture, can be recognized immediately at first sight."[27] Fuchs's disbelief that anyone could find fault with portraits of plants harkens back to a debate in the 1530s that he sustained with the physician Sabastian Montuus. The question under discussion was whether the visual medium could embody true botanical knowledge. Countering Fuchs's heated defense of such practices, Montuus maintained that illustrations displayed only accidents of form and were therefore not dependable sources of information. He alleged that the constant and unchanging qualities of plants—what he identified as their names and medical virtues—were much more reliable ways of identifying floral species.[28]

While Oviedo had no reason to be aware of the polemic between Fuchs and Montuus concerning plant illustration, its essence would have been all too familiar to him. Though couched in a new language, Montuus's distrust of illustration was one with an ancient lineage. No less than Pliny the Elder (23–79 CE) articulated similar concerns in the *Historia naturalis* regarding the dangers of attempting to reproduce images of plants. Pliny thought highly of artists like Timanthes, who took care to leave to the viewer's imagination those subjects which exceeded his abilities to depict. The need to recognize one's artistic limitations could have been one of the reasons for Pliny's unenthusiastic account regarding the achievements of Greek and Roman botanical illustration. Although Pliny seems to have admired advances in this field, he acknowledges the difficulties inherent in such an enterprise, writing,

> Among them, Crateuas, Dionysius and Metrodorus, adopted a most attractive method of description, though one which has done little more than prove the remarkable difficulties which attended it. It was their plan to delineate the various plants in colours, and then to add in writing a description of the properties which they possessed. Pictures, however, are very apt to mislead, and more particularly where such a number of tints is required, for the imitation of nature with any success; in addition to which, the diversity of copyists from the original paintings, and their comparative degrees of skill, add very considerably to the chances of losing the necessary degree of resemblance to the originals. And then, besides, it is not sufficient to delineate a plant as it appears at one period only, as it presents a different appearance at each of the four seasons of the year. . . . Hence it is that other writers have confined themselves to a verbal description of the

> plants. . . . Nor is this a kind of knowledge [of the properties and nature of plants] by any means difficult to obtain; at all events, so far as regards myself, with the exception of a very few, it has been my good fortune to examine them all, aided by the scientific researches of Antonius Castor, who in our time enjoyed the highest reputation for an intimate acquaintance with this branch of knowledge. I had the opportunity of visiting his garden, in which, though he had passed his hundredth year, he cultivated vast numbers of plants with the greatest care.[29]

Like Montuus, Pliny questions the ability of an illustration to fully express a plant's being, given the challenges of capturing the myriad forms it assumes in the course of its lifetime and from one season to the next. Prescient of the fate to befall the illustrated herbal centuries later, Pliny also warns of the errors in copyists' reproductions. Although Pliny does not deny the possible merits of plant images, he points to the care and attention that need to be exercised in this undertaking. The critical stance Pliny takes on this issue is related to his assumption that most readers had access to these plants in their natural environment, as did he. Pliny, as Oviedo centuries later, insists on the observation of a plant in situ; under such circumstances, he maintains, any given illustration will fall short of depicting the realities of an actual specimen.

Despite all the ways in which Oviedo follows in the footsteps of his revered predecessor, his decision to include plant illustration in the *Sumario* and the *Historia* (1535) is one that Pliny would not have endorsed with much enthusiasm. Nor is it likely that Pliny would have agreed with Oviedo regarding the value of adding color to woodcuts, a desire he expresses several times in the *Historia* (1535). Yet, as Oviedo demonstrates in other instances, he is acutely aware that his task does not afford him the same conveniences as those the Roman historian enjoyed. In his attempts to describe plant and animal species that readers would likely never see for themselves, Oviedo has no qualms about turning to images when words are insufficient. Regarding the species that Oviedo refers to as the welding tree, he writes in the *Sumario*, "It would be necessary to paint it so that what the tongue cannot express could be understood by way of sight."[30] Later on, in the Monserrat manuscript (1535–1549), Oviedo explains his reason for including an image of the cacao tree as follows: "Without a doubt, the eyes are a large part of the information of these things, and since they themselves cannot be seen or touched, the image of them is of much help to the pen."[31]

Like Fuchs, Oviedo insists on visual representation as a necessary counterpart to the limitations of the written word.

Carrillo Castillo and Myers maintain that Oviedo's woodcuts and drawings have no precursors in the illustrated print herbal nor any other literary genre that made use of visual representation, ranging from travel writing to encyclopedias.[32] While it is true that the developments reflected in Brunfels's *Herbarum* and Fuchs's *De historia* speak to advances in botanical illustration particular to Germany, however, other contemporary trends were not so geographically restricted. As we saw in Chapters 1 and 2, the study of the natural world occupied many minds in Europe during Oviedo's lifetime, including those of the Italian humanists with whom he engaged in conversation and epistolary correspondence, such as Giovanni Battista Ramusio (1485–1557) and Pietro Bembo (1470–1547). Artistic naturalism was in vogue in Italy, too: it was there that Oviedo likely saw engravings by Martin Schongauer (1448–1491), which were praised for their accurate portrayal of plants and animals. Oviedo also made the acquaintance of some of the foremost artists of the day—among them Leonardo da Vinci (1452–1519) and Andrea Mantegna (1430–1506)—who strove to reproduce nature in their art. Moreover, Oviedo was well read in authors and genres with visual traditions in manuscript and print form, although it is impossible to know which works he consulted in illustrated editions. In short, there is no reason to presume Oviedo's ignorance of print developments in northern Europe meant that his writings on the Indies did not respond to some of the same tendencies at work in these publications, and others like them. It is not a coincidence that the *Sumario* and the *Historia* (1535) were published roughly contemporaneously with Brunfels's and Fuchs's works: the innovative use of visual representation in all four books points to a shared cultural framework that informed these authors' decisions to make nature visible.

Experiencing nature, and representing that experience, were slowly becoming hallmarks of early modern natural history, both in Europe and abroad, in the Indies. Just as Brunfels's *Herbarum* and Fuchs's *De historia* ushered in a new era in the field of botany, so too did Oviedo's *Sumario* and *Historia* (1535) mark a novel approach to depicting visually the flora and fauna of the Americas. Despite the very different worlds these men inhabited, Oviedo, Fuchs, and the artist of the *Herbarum*, Hans Weiditz, confronted the same looming question: how does one most effectively illustrate nature? In response, each produced a series of plant images that differ radically from one another but that share the same claims of having been sketched from life.

Not content to let these visual displays speak for themselves, Fuchs and Oviedo explain and defend their use of graphic representation in written form. Thousands of miles away from one another, and separated by an immense ocean, these natural historians participated, each in his own way, in laying the foundation of the discipline of natural history.

Images of New World Flora in the *Sumario* (1526)

To say the woodcuts in the *Sumario* and the *Historia* (1535) depict what Oviedo saw is as much a true statement as a problematic one. Without question, his assurances of having experienced firsthand the phenomena about which he writes are the premise of his historiographical project. Regarding the illustrations of flora in particular, on several occasions Oviedo claims to have based his drawings on real-life models. Yet to take these statements at face value gets us no closer to understanding how Oviedo actively intervened to make this external reality fit the needs and objectives of his natural history. As the examples of Brunfels and Fuchs demonstrate, an avowal of fidelity to living plants can generate a myriad of unique visual interpretations. Far from copying blindly what he observes, the woodcuts of flora in Oviedo's published writings on the Indies represent an increasingly coherent visual program that he fashions deliberately and purposefully.

It has been the fate of the *Sumario* to exist in the shadow of its monumental successor, and its woodcuts are no exception. While its four illustrations pale in comparison to the dozens of woodcuts in the *Historia* (1535), they are a necessary point of departure for understanding Oviedo's initial approach to graphic representation and how it evolved over time. Above all, the woodcuts of the trivet tree (*árbol de las trébedes*) and the plantain leaf represent Oviedo's first attempts to give visual form to his encounters with the dense, tropical vegetation of the Caribbean coast of Central America, where he lived until 1523. Perhaps the most remarked feature of these illustrations is Oviedo's apparent inattention to them: although he places them alongside their corresponding written description, he makes no mention of them. The challenge to readers of puzzling out the relationship between word and image is unique to the *Sumario*; in the *Historia* (1535), Oviedo introduces explicit textual references to the illustrations. Beyond this similarity, however, the woodcuts of the trivet tree and plantain leaf share little in common. Embodying two wildly divergent attitudes toward illustrating New World nature, they do not fare equally in the later *Historia* (1535) and the Monserrat manuscript (1535–1549). Tracing these developments, from their earliest appearance in the *Sumario* to their subsequent iterations, is the focus of this next section.

Irreproducible Nature: Oviedo's Encounter with the Trivet Tree

The trivet tree stands apart from the other trees Oviedo explores in chapter 78 of the *Sumario* as the biggest he had ever seen. He introduces this unique specimen by way of an anecdote, recounting his chance encounter with it on official business on *tierra firme* in 1522.[33] He writes of his astonishment at the tree's massive size as well as its most salient feature—its three-pronged, above-ground root system. Oviedo's account of the tree, however, is undermined by the singular nature of his contact with it. Since this is the only tree of its kind he has come across, there is little information he can offer the reader apart from what he observed in passing. He cannot speak to the geographical distribution of the tree, the alimentary or medicinal uses of its leaves or bark, or variations in its growth and appearance, as he does with other flora in the *Sumario*. Perhaps most importantly, Oviedo is ignorant of the name of the species, so he invents one: the trivet tree, its root system reminding him of a trivet, an iron tripod used for cooking over a fire. To a large extent, Oviedo's encounter with the tree is individualized and irreproducible.

The accompanying woodcut is bound to these anecdotal circumstances (Figure 7). Oviedo portrays visually those features that, in his view, most defined his encounter with the tree: its three-pronged root system, rising out of the ground, and its massive trunk wrapped with vines, which Oviedo climbed to observe the surrounding landscape.[34] The impossibility of effectively conveying the enormity of the tree in such a restricted space becomes clear immediately in the lack of proportion between the tree's roots, trunk, and branches. Without a doubt, the image falls short of depicting the grandeur and majesty of the tree Oviedo describes.

Geographical limitations prevented Oviedo from including the episode of the trivet tree in the *Historia* (1535), dedicated primarily to flora and fauna on Hispaniola and the neighboring islands. It is only later, in the Monserrat manuscript, that this anecdote reappears. Its description is taken almost verbatim from the earlier passage in the *Sumario*, with the added qualification

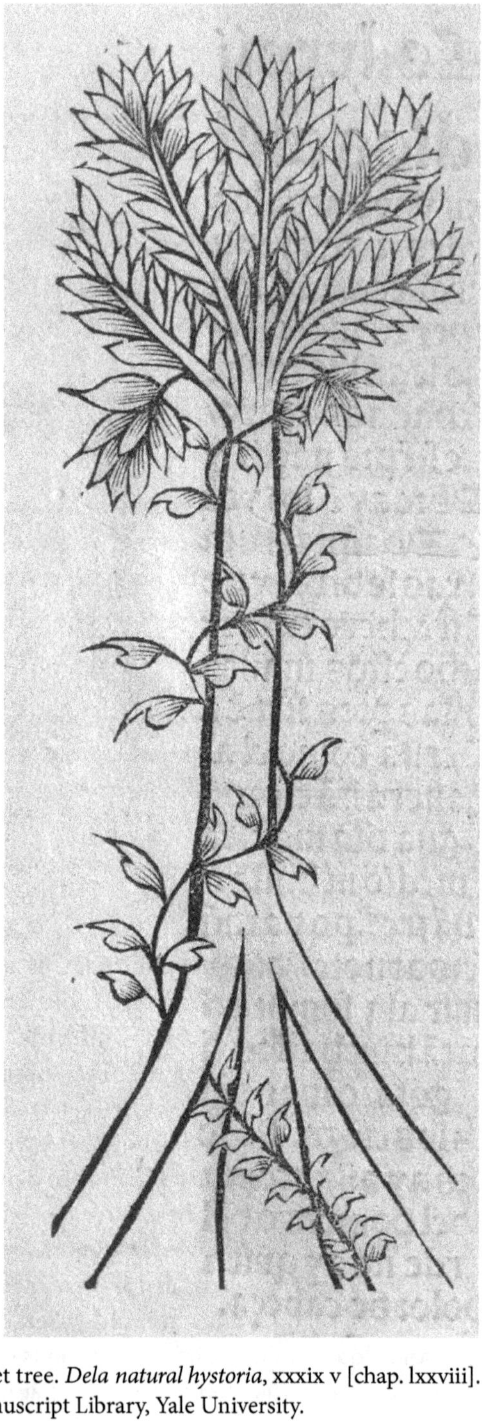

Figure 7. Trivet tree. *Dela natural hystoria*, xxxix v [chap. lxxviii]. Beinecke Rare Book and Manuscript Library, Yale University.

that since 1522 Oviedo had "seen many other, and much bigger [trees]" on *tierra firme*, which he refers to as ceiba.[35] The striking similarities between these two written accounts make conspicuous the absence of illustration from the manuscript. Indeed, the woodcut of the trivet tree is the only image from Oviedo's writings on the Indies that he does not recycle or modify for inclusion in a later text. In contrast, Oviedo adds the three remaining woodcuts from the *Sumario* to the *Historia* (1535) alongside twenty-six new illustrations, all of which he incorporates in one form or another into the Monserrat manuscript.

Oviedo does not explain why he removes the image of the trivet tree from his visual repertoire. Nevertheless, it is possible to speculate on several reasons for this decision. First, the fact that the illustration is linked to a fortuitous encounter Oviedo had with a particular tree makes it difficult for the woodcut to have meaning outside of this context. Despite the many years that have elapsed between the publication of the *Sumario* and the writing of the Monserrat manuscript (1535–1549), Oviedo is still unable to ascertain the tree's identity in the latter work; its namelessness continues to pose a challenge to further inquiries into a tree whose only known specimen is located in a specific time and place. So too, what most astounded Oviedo about the tree—its size—is a characteristic not easily conveyed visually. Not coincidentally, none of the other botanical or zoological illustrations in Oviedo's writings on the Indies attempt to reproduce such large phenomena in their entirety.[36] The absence of the woodcut of the trivet tree from the *Historia general* is an implicit acknowledgment of the futility of efforts to convey the monumental qualities of New World nature through visual representation. It is also one of the first indications of the extent to which, over time, Oviedo reevaluates the efficacy and function of graphic representations of nature.

A Case of Mistaken Identity: The New World Plantain

The singularity of the trivet tree contrasts sharply with the ubiquity of the plantain.[37] Although Oviedo maintains that the species is not endemic to the New World, it has multiplied to such an extent on the islands and mainland that "it is something that cannot be believed unless it is seen."[38] Its abundance allows Oviedo to provide a comprehensive account of the plant in the *Sumario* that acknowledges the difference and variety he has observed in its growth

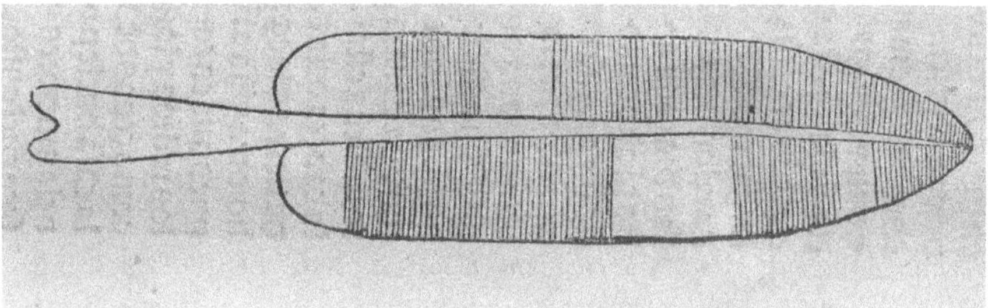

Figure 8. Plantain leaf. *Dela natural hystoria*, xlii v [chap. lxxx]. Beinecke Rare Book and Manuscript Library, Yale University. This woodcut is positioned vertically in the original source.

and appearance. Unlike the lone specimen of the trivet tree, about which Oviedo can offer little information, the pervasiveness of the plantain affords multiple avenues for learning about the plant and its various culinary uses.

The exhaustive verbal description of the plantain in the *Sumario* is juxtaposed with the selective nature of its visual representation, which features only its leaf (Figure 8). Oviedo implicitly establishes a parallel between the woodcut and its accompanying text: the long, delicate stem and the stacks of lines marking its blade put on display his observations regarding the fragility of the plantain's leaves, which rip apart in the breeze.[39] Oviedo's rationale for visually highlighting this aspect of the plant remains obscure, however, especially when considering that the protagonist of the written narrative is its fruit. It is only later, in the revisions Oviedo makes to his account of the plantain in the *Historia* (1535), that he clarifies the reasons that likely prompted this artistic decision. In so doing, he presents a more nuanced context from which to make sense of the illustration.

In the *Historia* (1535), for the first time, Oviedo casts doubt on whether the New World plantain truly qualifies as such, writing, "Truthfully it is not a tree nor a plantain, but rather some other kind of plant."[40] Despite his assuredness, his efforts to convince readers of the legitimacy of his claims is a task beset with difficulties. Since the "real" plantain originates in the Old World, Oviedo cannot resolve this case of mistaken identity through his normal routes of inquiry—namely, his and other colonists' firsthand experiences and recourse to indigenous informants. He must turn instead to the writings of other authorities who have knowledge of the species overseas: Pietro Martire d'Anghiera (1457–1526), the Italian travel writer Ludovico di Varthema (1470–1517), and Pliny the Elder.

While Oviedo is often dismissive and critical of Martire for having never set foot in the West Indies, the problem posed by the plantain reveals a different dimension to their literary relationship. Like Oviedo, Martire disagrees with the popular notion that the New World fruit is a plantain since it "greatly differs from the plantain, and is not related to it at all."[41] The Italian humanist's opinion is partially based on secondhand reports of the species from the Americas, but Oviedo is most interested in Martire's firsthand experiences in Egypt serving as an ambassador for the Catholic Monarchs. In *De orbe novo decades* (*Decades of the New World*), Martire recalls having eaten a fruit called *musa*, which in his view most closely resembles the New World plantain. Oviedo employs similar reports from Varthema's *Itinerario de Ludouico de Varthema Bolognese* (*The Itinerary of Ludovico di Varthema*) (Rome, 1510) to substantiate his assertion that the New World plantain is not the same species as its namesake.[42] These Italian intellectuals' observations of the plantain in the Middle East become a valuable counterpoint to Oviedo's impressions of its alleged equivalent in Spain's overseas colonies.

Having mined contemporary accounts for information on the plantain, Oviedo turns to Pliny's *Historia naturalis* for further evidence to support his hypothesis concerning the plant brought to the West Indies.[43] There is one particularly noteworthy difference between the plantain that Pliny describes and its supposed New World counterpart: the leaves. Oviedo includes no less than three anecdotes from Pliny's account illustrating how the tree offers protection from the elements, in one instance serving as a refuge from the wind and rain for as many as eighteen men.[44] The plantain that Pliny describes produces no fruit, so the respite it offers is its primary benefit. In contrast, the situation is the inverse with regard to the New World species: Oviedo writes that not only does it yield abundant fruit but the rips and tears in its leaves make it impossible for the plant to provide even a semblance of shade or protection. In fact, the constant dripping from the leaves gives the impression that it rains more beneath them![45]

Viewed from this perspective, the woodcut acquires meaning as a visual counterpart to Oviedo's argument concerning the true identity of the New World plant (Figure 9). The delicate leaf, displaying multiple rips and tears, is the conclusive evidence that the so-called New World plantain is, in fact, not a plantain at all. The changes Oviedo makes to the *Sumario*'s account of the species in the *Historia* (1535) cause readers to reformulate the relationship between word and illustration. Unlike the seemingly arbitrary nature

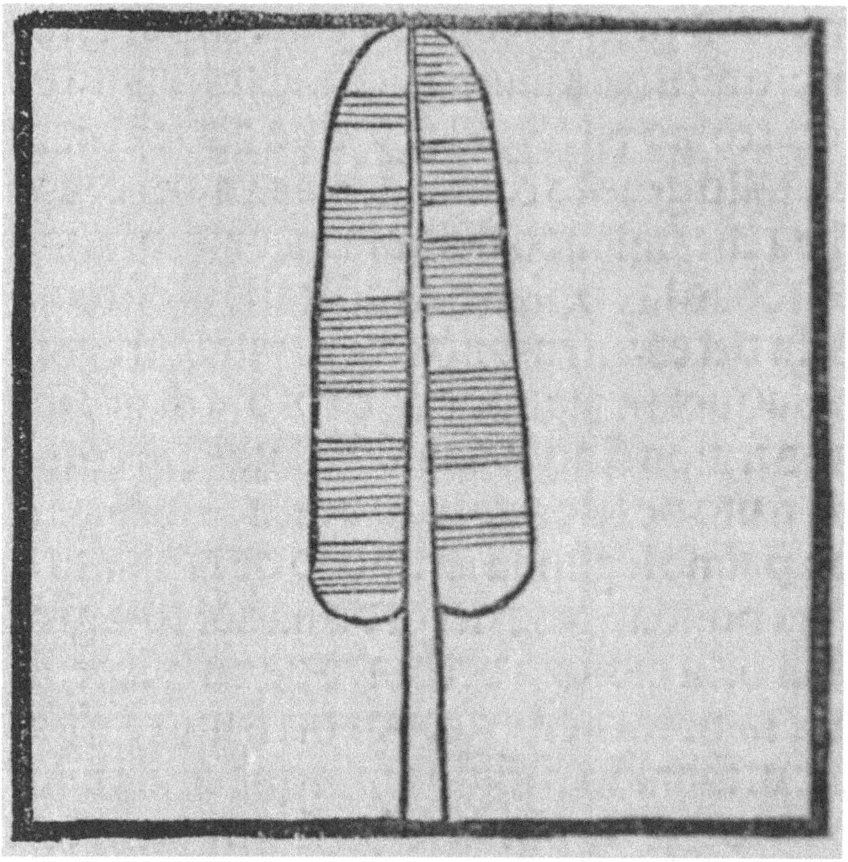

Figure 9. Plantain leaf. *Historia* (1535), lxxix r [bk. 8, chap. 1]. Beinecke Rare Book and Manuscript Library, Yale University.

of the image in the former, in the latter the woodcut becomes a powerful rhetorical display of the main site of divergence between the two plants.

Just as Oviedo comes to understand the *Sumario* as a "messenger or sign" of the *Historia general*, so too can the early woodcuts of the trivet tree and the New World plantain leaf be considered as precursors to the latter's illustrations.[46] The vast differences in the scale and scope of the two images attest to the varied criteria that Oviedo employs initially when determining how to render natural phenomena visually. The degree to which he modifies his treatment of these flora in the *Historia general*—removing the visual support for the trivet tree and strengthening the ties between the visual and verbal components of the description of the plantain—indicates how Oviedo continues to reflect on the role of visual and written language in conveying the botanical realities of the New World throughout his lifetime.

Nature Illustrations Explained: The Visual Program of the *Historia* (1535)

Despite its more limited geography, Oviedo's treatment of the natural world of the Indies in the *Historia* (1535) is more thorough than in the *Sumario*. Oviedo's continued experiences in the New World, as well as his access to the notes he had left behind when composing the *Sumario* in Spain, contribute to this disparity. More significant for this chapter, however, is the emergence in the *Historia* (1535) of an increasingly complex program of illustrating New World nature. Along with three woodcuts original to the *Sumario*, Oviedo includes twenty-six new ones. Roughly half of these, fifteen in total, depict New World flora, demonstrating their importance for him. The *Historia* (1535) marks the first time that Oviedo engages explicitly with these images. His comments about them point to some of the general tenets underlying his approach to illustrating New World nature.

By far the most commented-on aspect of the illustrations of flora in the *Historia* (1535) is their visual economy. Certainly, it comes as somewhat of a shock to see the vast reaches of New World nature reduced visually to a single leaf or a piece of fruit. Carrillo Castillo contends that this diminished frame of reference forces readers to rely on the written word for meaning by eliminating details that would "constitute excess and an invitation to intellectual curiosity and caprice."[47] Yet Oviedo offers no such rationale for these artistic decisions in his many remarks concerning the woodcuts in the *Historia* (1535). Quite to the contrary, Oviedo renders New World flora visually in ways that are designed to facilitate readers' understanding of the phenomenon in question. Be it in word or image, deciphering the natural world of the Indies for those who are less knowledgeable is Oviedo's most fundamental concern.

The abbreviated visual language that Oviedo employs in the *Historia* (1535) has the primary purpose of aiding in the identification of New World flora by drawing attention to their most noteworthy features. He explains his reason for including an image of the *guiabara* leaf (*Coccoloba uvifera*), writing, "It is so different and of such note among all other [leaves]" (Figure 10).[48] A

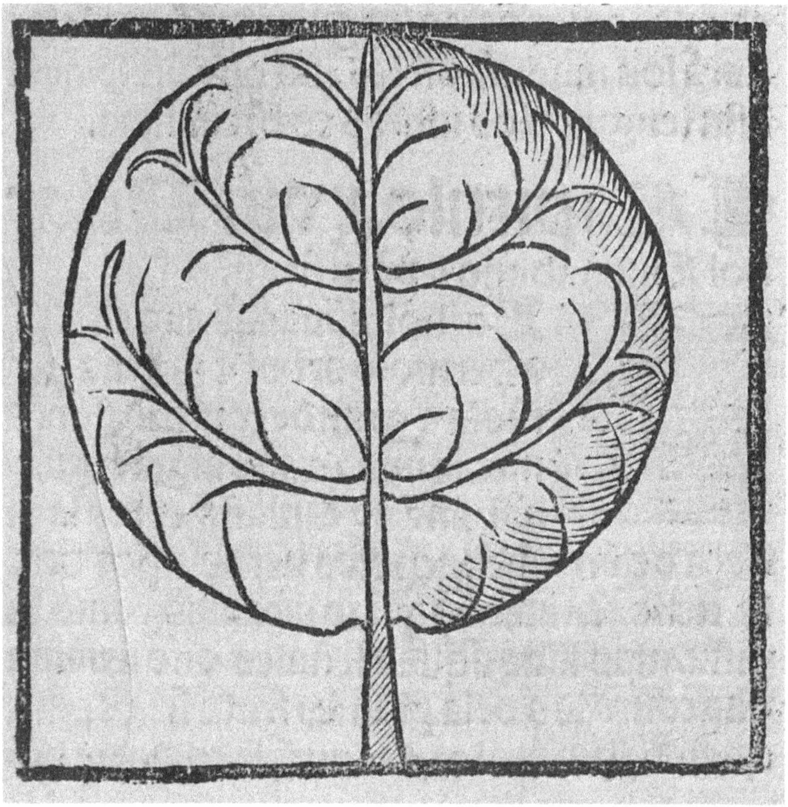

Figure 10. *Guiabara* leaf. *Historia* (1535), lxxxiii r [bk. 8, chap. 13]. Beinecke Rare Book and Manuscript Library, Yale University.

comparable logic informs the woodcuts of two different species of yuca, which Oviedo places next to each other in the text (Figures 11 and 12). Despite the many parallels between them, Oviedo observes one specific site of divergence: "There is a different generation of yuca that does not vary in its branches and fruit from the one I have just described, but rather in its leaf. Although it too consists of seven to nine lobes, it is of a different form."[49] Just as Oviedo visually represents the *guiabara* by its most singular trait, so too does he understand the leaves of the yuca as key to differentiating one plant from the other. The differences between the two are not purely aesthetic; Oviedo warns that one species of yuca is fatal when consumed raw. Oviedo uses these illustrations to draw attention to the plants' most salient features, which distinguish them from other flora.

Figure 11. Yuca leaf. *Historia* (1535), lxxiii r [bk. 7, chap. 2]. Beinecke Rare Book and Manuscript Library, Yale University.

Aspects of a plant's physical appearance acquire special significance for Oviedo when the plant is thought to have healing properties. In assigning meaning to the forms and patterns of nature, Oviedo draws on a tradition stretching back to Pliny's *Historia naturalis* in which a plant's morphological features were thought to reveal its curative abilities. The Doctrine of Signatures, as this theory was called, remained popular throughout the Middle Ages and into the Renaissance.[50] In contrast to Pliny, who ascribes to nature the qualities she incarnates, Oviedo credits God as the architect of these floral messages.[51] Nature is God's creation in need of deciphering; man must "read" plants to discover their utility for humankind. Oviedo attributes Spaniards' knowledge of the medicinal properties of the *perebecenuc* to the shape

Figure 12. Yuca leaf. *Historia* (1535), lxxiii r [bk. 7, chap. 2]. Beinecke Rare Book and Manuscript Library, Yale University.

of its leaf, which he describes as resembling a short iron spear. That they invite this comparison is, for Oviedo, a sign "as if to teach men that they cure the wounds of such spears."[52] He evokes the same imagery in the description of the artificial balsam (*bálsamo artificial*) plant, whose healing liquor remedies similar injuries.[53] Oviedo includes woodcuts of the leaves of both plants, availing himself of graphic representation to call attention to those visible aspects of nature that reveal divine mysteries (Figures 13 and 14).

Questions of visual representation in the *Historia* (1535) include not only what Oviedo chooses to depict but how. Only once in the work does Oviedo explicitly reference having consulted a real-life model for a sketch. Reflecting on his drawing of the artificial balsam leaf, Oviedo concludes: "This leaf [that I have drawn] is not badly shaped, as the plant in front of me is of the

Figure 13. *Perebecenuc* leaf. *Historia* (1535), xcviii r [bk. 11, chap. 5]. Beinecke Rare Book and Manuscript Library, Yale University.

same dimensions."[54] In reproducing rhetorically this pivotal moment of comparison, Oviedo attempts to assure readers of the sketch's accuracy (as well as that of the corresponding woodcut). One could imagine that Oviedo employed a comparable strategy when rendering other illustrations as well, given that he makes statements in the *Historia* (1535) regarding the degree to which they resemble their living counterparts.

That Oviedo sketched what he saw is precisely the kind of claim that traditionally has been accepted at face value. But such a statement should be examined carefully; the act of seeing, and reproducing visually what is seen, does not presuppose an indiscriminate gaze. Even though Weiditz and Fuchs employed artistic strategies similar to Oviedo's, the final products of their efforts are nothing alike. A case in point: whereas the woodcuts that grace

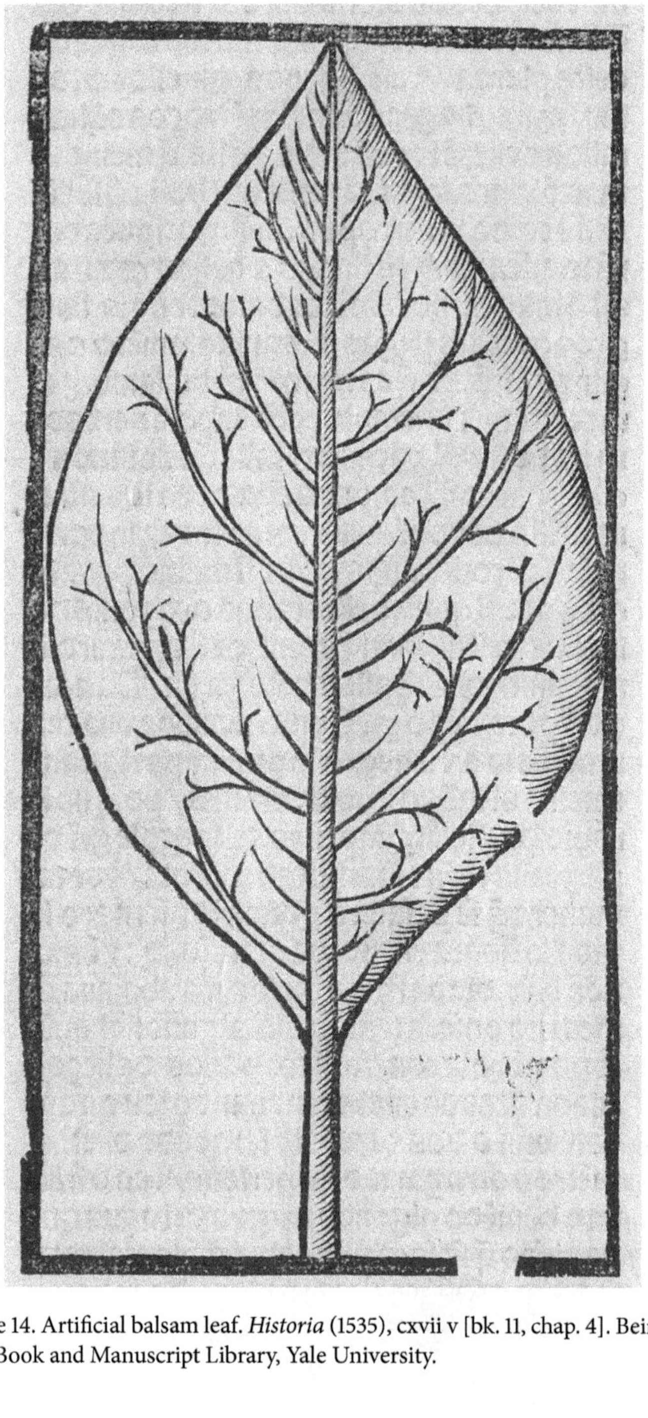

Figure 14. Artificial balsam leaf. *Historia* (1535), cxvii v [bk. 11, chap. 4]. Beinecke Rare Book and Manuscript Library, Yale University.

the pages of Brunfels's *Herbarum* depict the imperfections of individual plants, no blemishes mar the images of flora in the *Historia* (1535). These are flawless specimens, idealized versions of the phenomena they represent. Oviedo likely selected carefully and deliberately which specimens to sketch and almost certainly took artistic liberties when doing so.

Indeed, the woodcuts in the *Historia* (1535) demonstrate how Oviedo actively adjusted external reality to fit his needs. The trademark visual economy of these illustrations not only highlights plants' most distinctive features but also eliminates unnecessary details that could interfere with the process of recognition of a given species. Oviedo describes both kinds of yuca—the toxic as well as the harmless—as having between seven and nine leaves. Notably, this variation is not expressed in the corresponding images, whose stems have nine leaves each. This uniformity effectively eliminates the danger of readers' misinterpreting arbitrary visual decisions as true differences between the two plants.

This systematic approach to many of the woodcuts in the *Historia* (1535) is also visible in their generic nature. Oviedo acknowledges in writing variance in the size and shape of a given species; visually, however, he presents what is meant to be a standardized rendering of the phenomenon in question. Typical of the entries on flora in the *Historia* (1535) are Oviedo's comments regarding the illustration of the *perebecenuc* leaf: "The leaf of this herb is of the form and figure that is drawn here, although some are bigger and some smaller."[55] By qualifying that the leaf depicted is both smaller than some real-life specimens and larger than others, effectively the woodcut functions as a kind of average or norm for the species.

These strategies that Oviedo uses to illustrate New World nature can be equated with Leonardo da Vinci's work with synthetic naturalism, represented most famously by his ideally proportioned sketch of a horse. To compose the drawing, the Florentine artist made a series of sketches of various horses; from these, he determined the best features of each model and copied them to create a composite illustration.[56] Oviedo, too, creates his own idealized versions of the plants he represents visually, taking care not to portray details particular to individual specimens nor the extremes and variety inherent in the species. These illustrations are their own kind of composite image, in that they showcase what is most typical.

Oviedo is well served by illustration in his efforts to render visually the form of a plant and its relative proportions. He struggles, however, to communicate other features via this medium that he holds equally important.

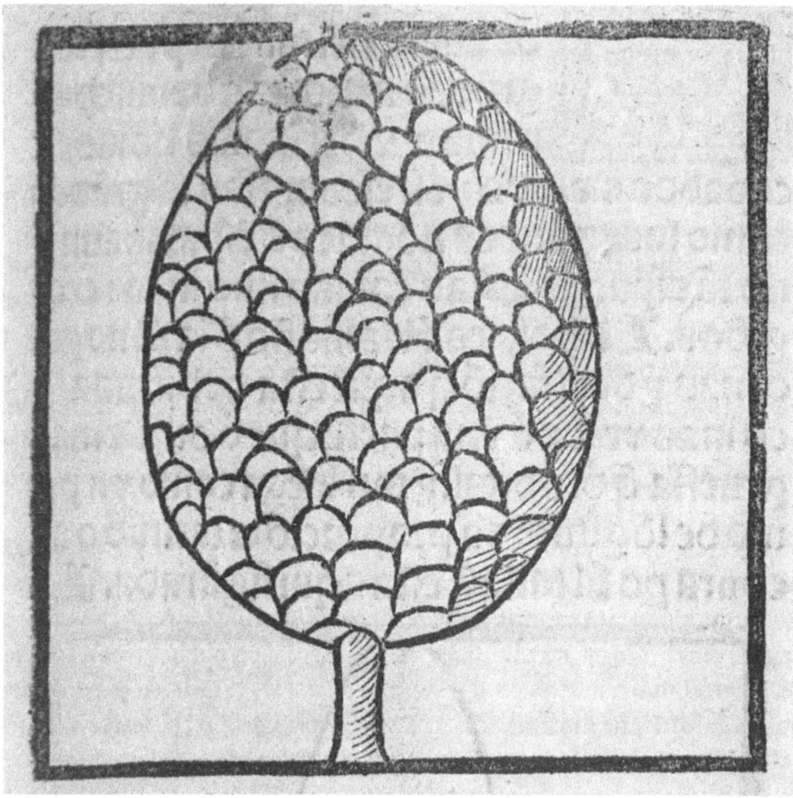

Figure 15. *Guanábana* fruit. *Historia* (1535), lxxxiiii r [bk. 8, chap. 17]. Beinecke Rare Book and Manuscript Library, Yale University.

Had he been able to shoulder the cost, Oviedo would have gladly imparted color to the woodcuts in the *Historia* (1535); he laments its absence in the cases of the pineapple and the welding tree as essential to their identity.[57] So too, without color Oviedo is unable to distinguish between the anon and the *guanábana*; he maintains that these two fruits are alike in every way, save that the latter is green and the former yellow. In lieu of being able to portray this distinction visually, Oviedo represents them both with the same woodcut: "I will not repeat the [image of the] fruit [the anon] because it has the same drawing as the *guanábana* above" (Figure 15).[58]

Occasionally, Oviedo compensates for this lack of color through his use of shading. His comments regarding such features indicate that he intended these details to be reproduced in the corresponding woodcuts as hatch marks.

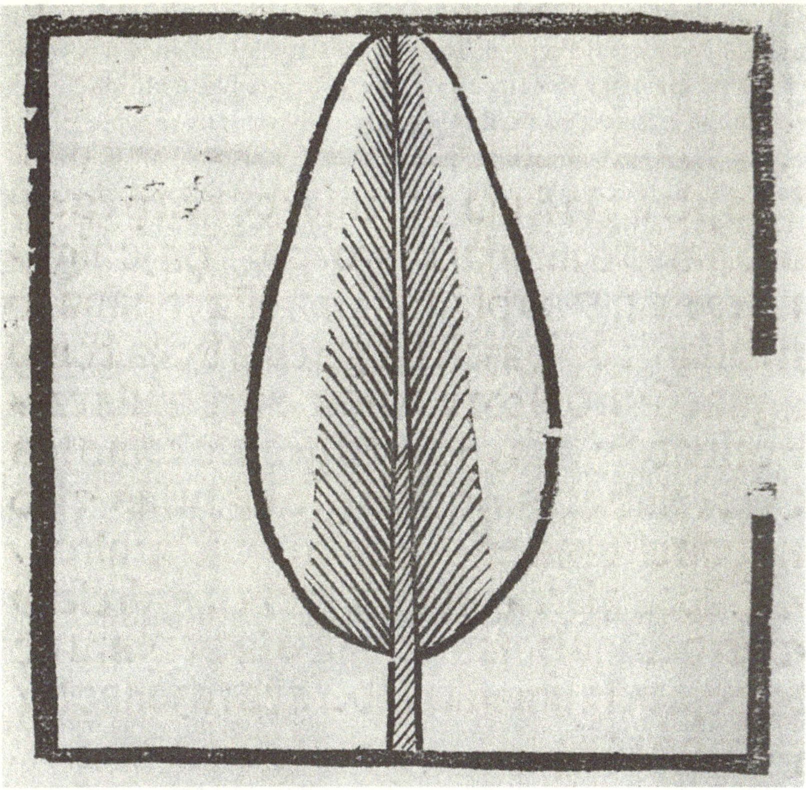

Figure 16. Mamey leaf. *Historia* (1535), lxxxiiii v [bk. 8, chap. 20]. Beinecke Rare Book and Manuscript Library, Yale University.

Oviedo interprets the parallel lines marking the outside edges of the *perebecenuc* leaf as follows: "That hue or shade on the tips of the leaves, as they are drawn, should be understood as purple."[59] The shaded section of the leaf provides a point of contrast to its lighter areas, which Oviedo describes as green and white (Figure 13).[60] Similarly, the hatch marks surrounding the rib of the mamey leaf (*Pouteria sapota*) visually suggest a contrast between two different shades of green (Figure 16).[61] In these instances, hatching presents a partial solution to not having recourse to color.

The woodcuts of flora in the *Sumario* and the *Historia* (1535) are one of Oviedo's most enduring claims to fame. At the same time as they celebrate the biological possibilities of Spain's overseas colonies, they participate in the shift taking place within Europe's scientific community of increasingly privileging botanical illustration. It is unlikely that Oviedo was acquainted with the works of Brunfels and Fuchs, nor they his; nonetheless, that each developed a visual program for documenting flora concurrently with one another points to the existence of a diffuse early scientific culture in which they all took part. Far from being a mere peer of these German botanists, Oviedo is a pioneer in his own right, facing challenges that Weiditz and Fuchs never encountered. In the *Sumario*, Oviedo begins to experiment with various visual strategies to make understood a reality that his readers will likely never experience firsthand. In the *Historia* (1535), he further refines this approach; the visual economy that epitomizes Oviedo's portrayal of New World flora reflects a deliberate and systematic engagement with issues of representation. By interrogating the seemingly objective claims of artistic mimesis at work in these two publications, we are better able to appreciate the degree to which Oviedo continues to reflect on the role of illustration as part of his constantly evolving project of writing a history of the New World.

The Fate of the Woodcuts in the Monserrat Manuscript

The images in the *Sumario* and the *Historia* (1535) constitute a preliminary step in a visual program that Oviedo envisioned as culminating in a three-part history of the Indies.[62] Starting around 1535, Oviedo began to add to and edit the nineteen books of the *Historia general* that had already been published. During the subsequent fourteen years, until 1549, Oviedo also wrote parts 2 and 3 of the work, which addressed the ever-expanding territories of Spain's conquest and colonization. Ultimately, this newly revised and augmented *Historia general* was not published in Oviedo's lifetime, and large portions of the Monserrat manuscript have since been lost.[63] Although copies survive of some of these lost originals in the so-called Truxillos manuscript from the sixteenth century and the Spanish historian Juan Bautista Muñoz's eighteenth-century copies, others disappeared before they could be transcribed.[64] Given these circumstances, our knowledge of the illustrated material in the Monserrat manuscript is incomplete.[65] That being said, the still-extant books on flora indicate that Oviedo planned not only to recycle and/or revise the illustrations from the *Historia* (1535) but also to add more visual material: the manuscript of *Book 7* has two new hand-drawn illustrations, and *Book 9* and *Book 11* contain three and four, respectively. A comparison of the images of flora in the *Historia* (1535) and the Monserrat manuscript demonstrates the continuities and differences between these two phases of Oviedo's efforts to illustrate New World nature.

Perhaps what most set apart the illustrations of flora in the Monserrat manuscript from their counterparts in the *Historia* (1535) are their dimensions and placement in the text. The pineapple, whose woodcut occupies a little more than a quarter of a folio in the *Historia* (1535), increases in size and prominence in the Monserrat manuscript (Figure 17). Placed squarely in the center of the page, the drawing of the fruit was clearly composed prior to Oviedo adding the accompanying written description. The same is true of the manuscript illustrations of the artificial balsam leaf, the *perebecenuc* leaf, the lily, the *perorica*, and the *coigaraca* (Figures 21, 20, 24, 25, and 26). In and of themselves, however, it is difficult to reach any definitive conclusions about these changes. Not only do the images correspond to different artistic mediums—woodcuts in the *Historia* (1535) and drawings in the Monserrat manuscript—but since no prior manuscript copy exists of Oviedo's printed works on the Indies, it is impossible to know the extent to which his original drawings were modified for print. It is conceivable that Oviedo treated them similarly to those in the Monserrat manuscript before they were subjected to the limitations of the press.

Let us turn our attention, then, to other changes that took place in Oviedo's visual program over time. In the Monserrat manuscript, Oviedo modifies almost all the images that had previously featured in the *Historia* (1535). In his two drawings of yuca leaves, Oviedo relaxes the stems and adds a second set of leaves, virtually identical to the first except newly sprouted (Figures 18 and 19). In the margins next to the illustrations, he includes two identical notes to the printer: "It is done, but not like this."[66] It is likely that Oviedo expected these revisions to be incorporated into a future print edition of the work.[67] Although he does not indicate the reason for these artistic interventions, it is undeniable that they resulted in a decidedly more lifelike final product.

The drawing of the *perebecenuc* in the Monserrat manuscript also displays an additional leaf (Figure 20). Oviedo's instructions to the printer here are more explicit than those concerning the images of the yuca, although the reason for his request is unclear: "It is to be done in the likeness of this one."[68] In contrast, the changes Oviedo makes to the drawing of the artificial balsam leaf can be attributed to his desire to clarify the relationship between the visual and verbal elements of the text. In the *Historia* (1535), Oviedo describes the shape of the leaf as resembling an iron spear, that is, sharp on both ends, but the accompanying woodcut does not follow suit.[69] It

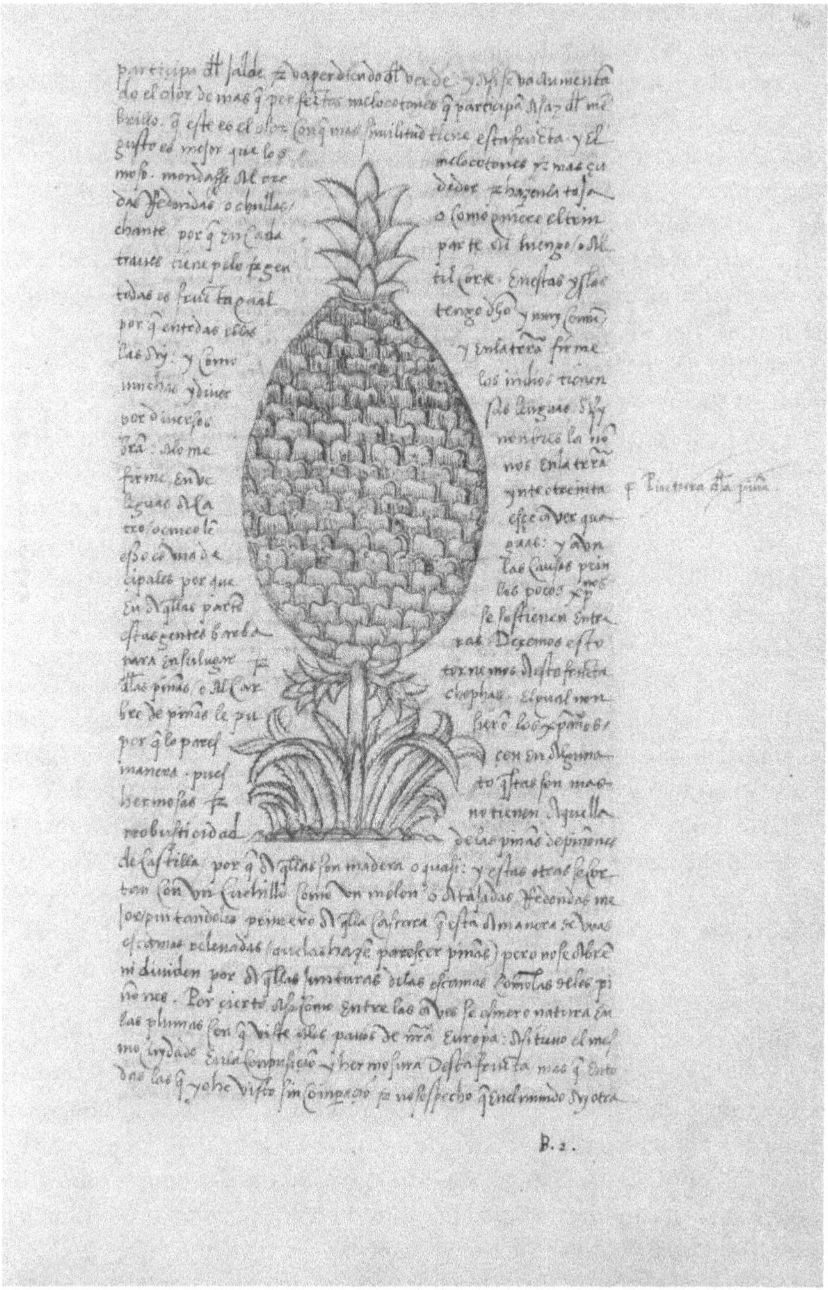

Figure 17. Pineapple. *Historia general*, MSS HM 177, 46r [bk. 7, chap. 8]. The Huntington Library, San Marino, California.

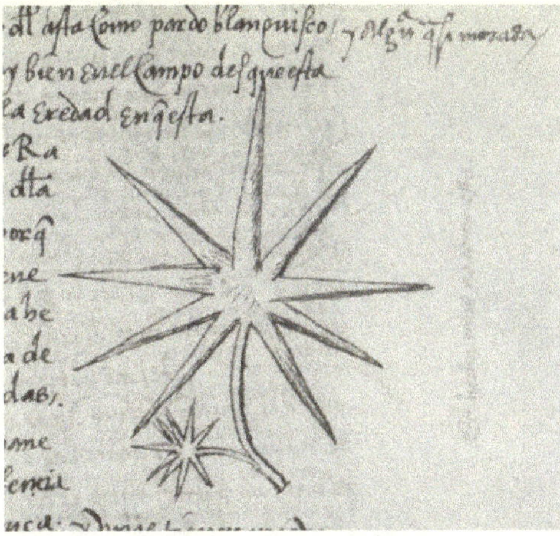

Figure 18. Yuca leaf. *Historia general*, MSS HM 177, 37r [bk. 7, chap. 2]. The Huntington Library, San Marino, California.

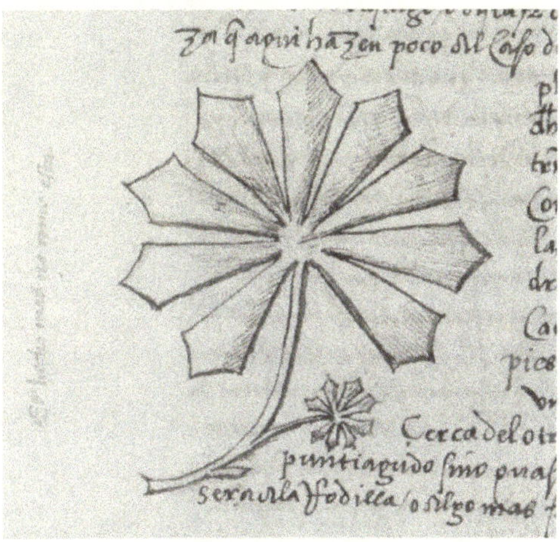

Figure 19. Yuca leaf. *Historia general*, MSS HM 177, 37r [bk. 7, chap. 2]. The Huntington Library, San Marino, California.

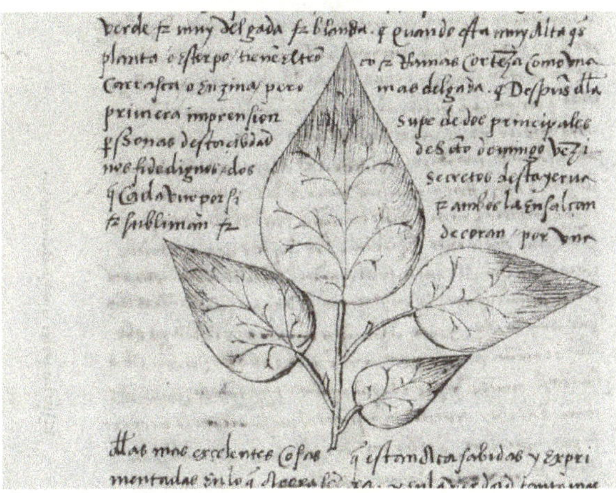

Figure 20. *Perebecenuc* leaf. *Historia general*, MSS HM 177, 76v [bk. 11, chap. 5]. The Huntington Library, San Marino, California.

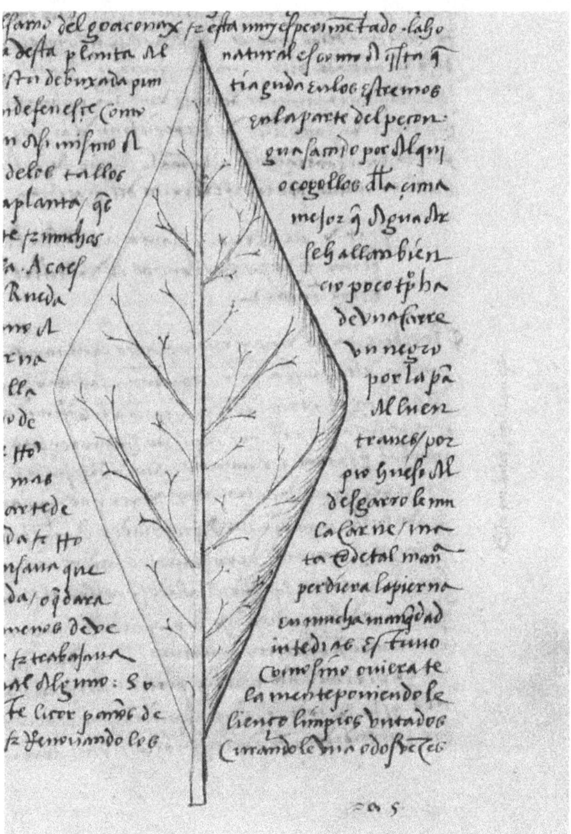

Figure 21. Artificial balsam leaf. *Historia general*, MSS HM 177, 74r [bk. 11, chap. 4]. The Huntington Library, San Marino, California.

is only later, in the Monserrat manuscript, that Oviedo changes the gentle curved lines of the bottom of the leaf into sharp angles, thereby strengthening the correspondence between image and word (Figure 21).

In addition to the woodcuts of flora that Oviedo recycles from the *Historia* (1535) in the Monserrat manuscript, he creates eight new images: a maguey plant (agave) (*Book 7*), a coconut tree (*Book 9*), a *capera* pod (*Book 9*), cinnamon (*Book 9*), the herb *y* (*Book 11*), a lily (*Book 11*), a *perorica* plant (*Book 11*), and a *coigaraca* plant (*Book 11*).[70] With these, Oviedo begins to experiment with other forms of representation, acquiring an interest in depicting a plant's internal features as well as its outward form. In depicting the *capera*,

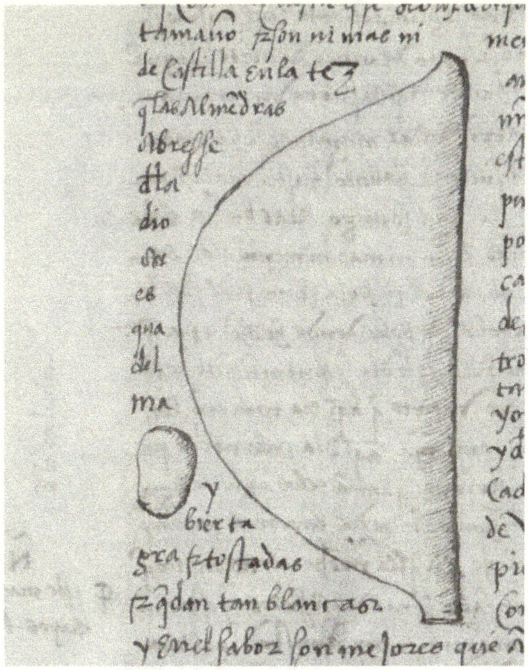

Figure 22. *Capera*. *Historia general*, MSS HM 177, 66v [bk. 9, chap. 26]. The Huntington Library, San Marino, California.

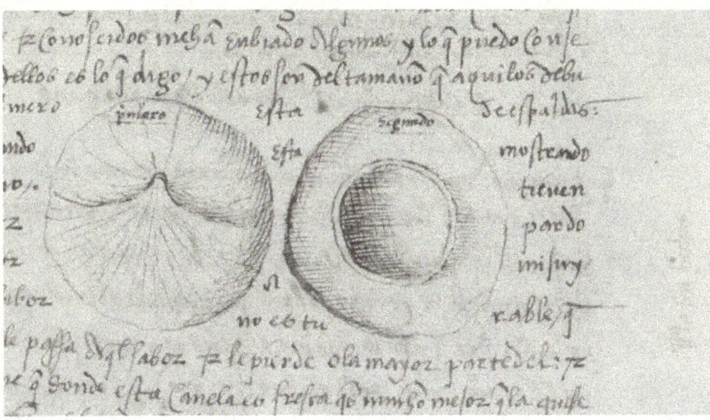

Figure 23. *Canela*. *Historia general*, MSS HM 177, 68r [bk. 9, chap. 31]. The Huntington Library, San Marino, California.

Figure 24. Lily. *Historia general*, MSS HM 177, 78r [bk. 11, chap. 7]. The Huntington Library, San Marino, California.

Figure 25. *Perorica. Historia general*, MSS HM 177, 78v [bk. 11, chap. 8]. The Huntington Library, San Marino, California.

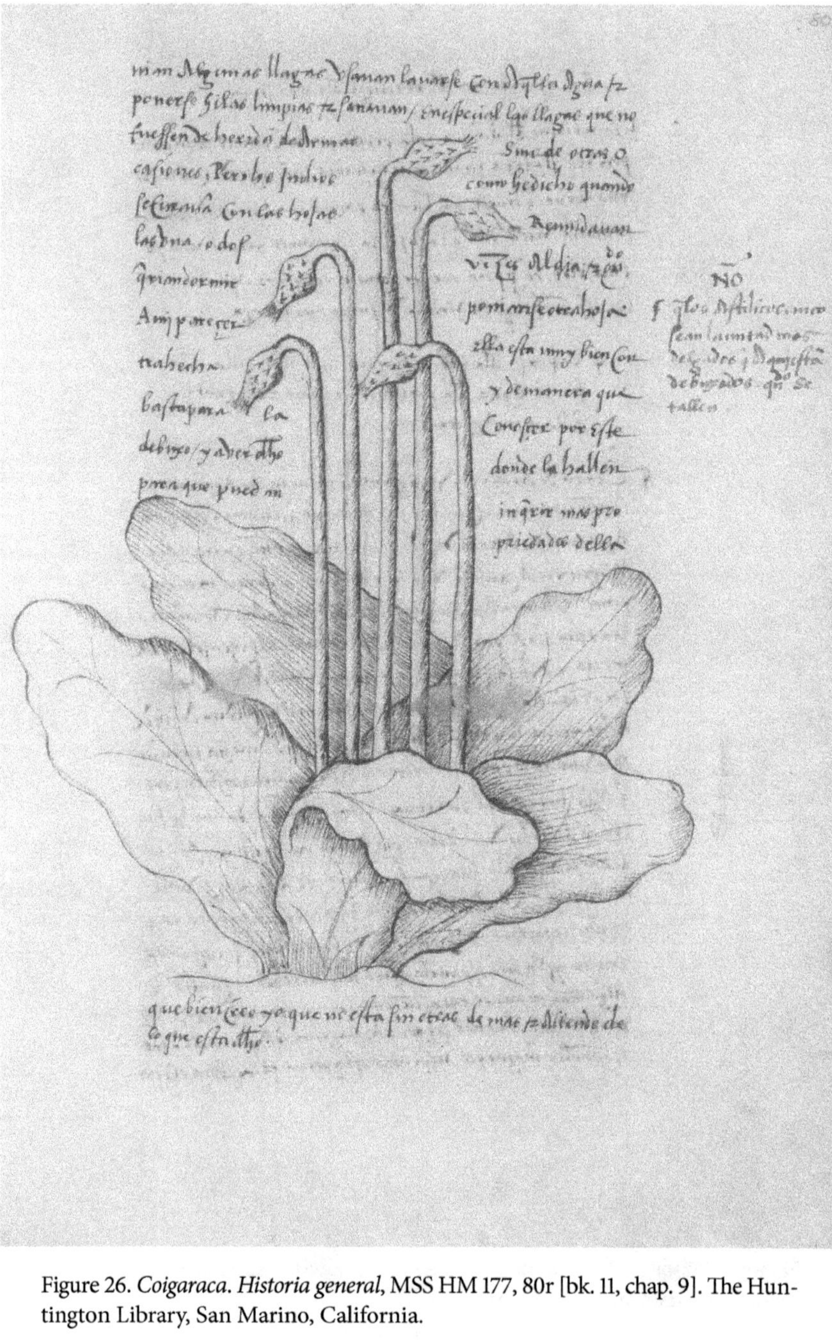

Figure 26. *Coigaraca. Historia general*, MSS HM 177, 80r [bk. 11, chap. 9]. The Huntington Library, San Marino, California.

Oviedo draws the hard shell of the almond-like nut alongside one of the four small edible seeds it contains, claiming that both are true to size (Figure 22).[71] So too, Oviedo captures visually the outside of a cinnamon pod (*canela*) as well as its empty interior (Figure 23). (Oviedo received the specimen from an acquaintance and so did not know what its seed looked like.)[72] These images suggest that, after the publication of the *Historia* (1535), Oviedo developed a more discerning sense of plants and their various parts—beyond just their leaves and fruits—as essential markers of identity and function.

Still other new drawings in the Monserrat manuscript—that of the lily, *perorica*, and the *coigaraca*—portray whole specimens (Figures 24, 25, and 26). In their completeness, these illustrations recall the trivet tree in the *Sumario* and the three woodcuts of cacti in the *Historia* (1535), the latter a topic of the next chapter. Nevertheless, they mark a novel phase in Oviedo's visual repertoire, of note for the new features he deems necessary to include: the flowers of the lily and the roots of the *coigaraca* and *perorica*.

Increasingly, then, in the course of his career Oviedo seeks visual comprehensiveness. In these final drawings, he reevaluates his earlier approach to plant illustration: no longer are species depicted through a representative selection of their parts but rather as complete specimens. By considering other alternatives to illustrating New World nature, Oviedo continues to search for the most effective way of harnessing the power of the visual medium. Yet, despite the inarguable differences between the woodcuts in the *Sumario* and the *Historia* (1535) and the drawings in the Monserrat manuscript, underlying all these efforts is the same fundamental concern: making New World nature understood.

CHAPTER 6

(Natural) Limitations

Monopolies and Monstrosities in the *Historia* (1535)

One of the major challenges Oviedo undertakes as a natural historian of the Indies is effectively and accurately communicating his knowledge to his readers. Particularly problematic in this respect are several plant and animal species that defy Oviedo's ability to understand, try as he might. The epistemological obstacles posed by the balsam, cactus, and iguana in the *Historia* (1535) vary in their details; collectively, however, they showcase Oviedo's reaction to insurmountable barriers, both natural and man-made, to his knowledge. He is unable to describe the balsam in as much detail as he would like because of the conditions of the Spanish entrepreneur Antonio de Villasante's monopoly over the production and sale of the plant's healing liquor. The cactus and iguana, too, pose unique dilemmas: the highly irregular forms of the cactus and the enigmatic behavior of the iguana puzzle Oviedo to no end. Despite the vast differences in Oviedo's literary and visual treatment of these three species, when his descriptions of them are viewed one alongside the other they offer a powerful commentary on the limitations of his early efforts to apprehend New World nature.

Antonio de Villasante and the Commercialization of New World Balsam

The year 1526 was a propitious one for both Oviedo and Antonio de Villasante. Independently of one another, the two had journeyed to Spain from their home on the island of Hispaniola—Oviedo in 1523 and Villasante several years later—to seek the royal patronage of Charles V for their respective proposals. Although each of them looked to New World flora and fauna to further his ambitions, the fruits of their time at court took vastly different forms: in the dedicatory to the *Sumario* (1526), Oviedo expresses his appreciation at the young monarch's insistence that the work be published, so that "these great and marvelous and new things [become] known to all men."[1] In contrast to the spirit of sharing so integral to Oviedo's project, Villasante had quite another objective in mind in his petitions to the Crown. Claiming to have discovered a tree the astute entrepreneur referred to as *bálsamo* or balsam, as the Egyptian remedy was known, Villasante sought to protect his stake in the plant and maximize his financial gain. In 1526, just as Oviedo saw the *Sumario* come off the press, Charles V confirmed the monopoly that the admiral Diego Colón had previously conceded to Villasante for exclusive trade in the medicinal balm. That the king endorsed both these ventures—Oviedo's *Sumario* and Villasante's commercial dealings in *bálsamo*—points to his desire to know more about his overseas possessions as well as to exploit their resources.

In practice, however, these two aims were not always complementary. For his part, Oviedo understood the legislation issued to protect Villasante's business dealings as infringing directly on his efforts to generate knowledge about the natural resources of the Indies. Others, too, protested the Crown's decision: almost immediately upon Villasante's return to Santo Domingo, residents of Hispaniola as well as the neighboring island of San Juan (Puerto Rico) began to grumble about the conditions of the monopoly, which placed the harvesting, distribution, and sale of the tree entirely in the hands of Villasante and his associates. Among these discontented voices were those who questioned the true likeness between the *bálsamo* and its Old World namesake, the Egyptian balsam. They suggested that the decision to name the *bálsamo* as such was a marketing strategy that put the health and well-being of the Crown's subjects at risk. Oviedo's comments regarding the *bálsamo* in the *Historia* (1535) emerge from similar concerns, as well as others, particular to his task as a natural historian of the Indies. Characterized by silence and evasion, Oviedo's account of Villasante's prized tree stands in stark opposition to his treatment of other, similar medicinal plants, such as the artificial balsam (*bálsamo artificial*) and the *perebecenuc*. Oviedo holds the ability to share and exchange information about the natural world of the Indies as vital for the future prosperity of Spain; in the *Historia* (1535), he demonstrates how Villasante's monopoly over the balsam compromises this undertaking.

Villasante's decision to call this New World tree *bálsamo* was most certainly a deliberate one, made in the hopes of ensuring its commercial success. In a report to the Crown about the plant and its medicinal properties—written around 1528—Villasante references several different indigenous names for the species, including *boni*, *guacuñaso*, and *canaguey*.[2] Ultimately, however, he rejected these alternatives in favor of a name that was sure to resonate with potential customers in Europe—balsam. A medicinal remedy from Matariyya, Egypt, the Old World balsam was prized for its healing properties by classical and contemporary authors alike. The liquor produced from the tree was said to cure a host of maladies: it healed wounds; counteracted snake bites and hemlock poisoning; diminished cataracts; and alleviated respiratory ailments such as persistent coughing, pneumonia, and asthma.[3] Its value as a remedy was matched by its commercial worth: Pliny the Elder mentions the high price that the aromatic tree resin fetched at the market in Rome.[4] In early modern Europe, as in Roman times, balsam continued to be a highly sought-after luxury commodity, albeit one increasingly difficult to acquire. By the beginning of the sixteenth century, there were no more living balsam trees growing in the garden at Matariyya, resulting in a situation where demand greatly outpaced supply.[5] Villasante's decision to style the New World tree after this Old World counterpart allowed him to take advantage of the prestige of its namesake and the deficit in its supply.

No wonder Villasante's proposal was endorsed so enthusiastically by Charles V. The idea of a New World balsam appealed to the Spanish Crown as a means to benefit from an already established global trade at a moment that could not have been more opportune.[6] In addition to guaranteeing the entrepreneur a third of the profits from the sale of balsam, as well as any other drug he was interested in marketing, the king awarded him a host of other privileges as well. These included tax and labor incentives and the positions of local magistrate (*alcalde*) and alderman of the city council of Santo Domingo (*regidor*).[7]

The news of Villasante's success at the royal court met with much consternation among the residents of Hispaniola, and they did not hesitate to relay their concerns to the Spanish monarch. One particularly vocal opponent of the commercial arrangement was *licenciado* Barreda, a physician in Santo Domingo and Oviedo's longtime acquaintance. In a letter written to Charles V in 1528, Barreda disagrees with physicians overseas who "approved . . . as balsam the liquor that Villasante took there [to Spain]."[8] Were His Majesty's subjects made to buy and administer the falsely advertised liquor, he continues, it could play havoc with their health and financial well-being.[9] To remedy the situation, Barreda urges the king to consult New World physicians, who have "long experience and much knowledge" of the plant and its healing liquor.[10] Barreda does not deny the medicinal properties of the so-called balsam. Rather, he takes issue with the misplaced trust in medical professionals in Spain to verify the properties of New World species. Barreda insists on the importance of local knowledge, without which attempts to prescribe uses for foreign remedies could have dire consequences.[11]

Licenciado Barreda was not the only New World resident to voice concerns about Villasante's commercial endeavors. A colonist of San Juan (Puerto Rico), Pedro Sánchez de Valtierra, also wrote to the king in 1528, requesting that the residents of the island be exempt from the prohibition of harvesting balsam.[12] Valtierra expresses his frustration at having to buy the medicinal remedy from Villasante when the plant grows abundantly on the island. As a potential solution, he proposes that residents be given leave to pick balsam for noncommercial purposes.[13]

Unfortunately, Barreda's and Valtierra's requests fell on deaf ears: the Crown had already issued a series of royal charters to protect Villasante's monopoly over the making and selling of balsam.[14] Several years later, in 1530, two other royal charters reiterated the Crown's stance: one reminded the bishop of Santo Domingo and La Concepción that residents of Hispaniola should not make ointment from the tree and the other cautioned ships' captains against accepting balsam from anyone other than Villasante and his associates.[15] Still more legislation emerged that same year seeking to suppress harmful opinions about the tree and its efficacy as a remedy. Queen Isabella of Portugal, acting as regent of Spain, ruled that physicians should not publish medical opinions about the balsam without first submitting them to rigorous testing and reporting to the local magistrate any findings unfavorable to the liquor.[16] Ultimately, all these measures were designed to ensure the commercial success of the balsam. Not only did they place the production and distribution of the healing ointment exclusively in Villasante's hands but they also attempted to control dissemination of knowledge about the tree.[17]

Like *licenciado* Barreda and other island residents, Oviedo also objects to the Crown's handling of Villasante's proposal. Despite the entrepreneur's ties to Santo Domingo, as well as the fact that he was present in Charles V's court at roughly the same time as Oviedo, in the mid–1520s, Oviedo does not acknowledge having met Villasante until after 1535, in the Monserrat manuscript (1535–1549).[18] Given that Oviedo records the entrepreneur's death in the *Historia* (1535), however, it is likely that this meeting happened much earlier. In any case, Oviedo was well positioned to learn the details of this business dealing and the controversy it elicited: he knew the admiral Diego Colón, who had first granted Villasante the rights to trade balsam, as well as *licenciado* Barreda.[19] Oviedo holds the physician in high esteem, referring to him as a skilled doctor.[20] Although Oviedo does not mention having discussed this topic with Barreda, in the *Historia* (1535) he seconds the physician's opinion that "balsam" is a misnomer for the New World species.[21]

Whereas Barreda employs Dioscorides to compare the tree to its namesake, however, Oviedo does not substantiate his claims that the two are different species. He states simply, "Of the true balsam, Pliny [writes] in book 12, chapter 26, and many other authors have written, so there is no need to discuss it here, since the effects of the good balsam are appropriate to other things at odds with those that are cured with this artificial liquor or what some want to cure."[22] His refusal to elaborate on the subject is unusual, considering how often he weighs in on similar issues as an expert on New World nature. Recall that Oviedo chastises those who dub as "terebinth" a species of trees found in the Indies, employing evidence from Pliny's *Historia naturalis* to demonstrate the differences between the New World plant and the one the Roman historian describes.[23] Likewise, he enlists the help of several classical and contemporary historians to oppose popular opinion regarding the identity of the New World plantain.[24] That Oviedo does not take a comparable approach to the confusion surrounding the New World balsam is an early indication of how radically his treatment of the plant differs from that of other flora of the Indies.

Oviedo makes no secret of his opposition to the legal protections afforded to Villasante's enterprise, leading Enrique Álvarez López to remark on his aversion to "any monopoly or private concession like that requested or conceded to one Villasante."[25] Nonetheless, in general this episode in the *Historia* (1535) has not received much scrutiny: Ernst Schäfer's and Esteban Mira Caballos's articles on the New World balsam do not explore Oviedo's stance on the subject, and Antonio Barrera-Osorio touches only briefly on the topic.[26] Yet such cursory attention to Oviedo's opinions on Villasante's monopoly overlooks the deep-seated misgivings he expresses concerning the dangers it presents for the exchange and circulation of knowledge about New World nature. While normally Oviedo strives to communicate as clearly as possible with readers, his account of the balsam reveals a different side to him, one in which he privileges silence and ambiguity over concrete answers. By erecting visual and verbal barriers to knowledge of the New World balsam, Oviedo calls attention to how the privatization of natural resources obstructs attempts to learn more about them.

Silent Objections: The New World Balsam in the *Historia* (1535)

Oviedo opens his account of the New World balsam in *Book 10* by questioning the plant's identity and Villasante's rights to its medicinal liquor. He insists that the tree, which natives refer to as *goaconax*, is not balsam. Nevertheless, the plant is so widely known by this name that Oviedo refers to it as such.[27] He subtly undermines Villasante's attempts to lay claim to the "secret" of the balsam's healing ointment by relating two conflicting versions of its discovery.[28] According to Villasante, his indigenous wife shared the remedy with him; a second rumor credits this finding to the Italian doctor and philosopher Codro, who died shortly thereafter.[29] Ultimately, however, knowledge of the preparation of the healing liquor was attributed to Villasante, for which he was "rewarded . . . by the imperial majesty of the emperor our king."[30] By casting doubt on Villasante's claims to ownership over the balsam's ointment, Oviedo implicitly calls into question the entrepreneur's stake in its profits.

While Oviedo's eyewitness testimony is the protagonist of his natural history, he is at pains to emphasize his unfamiliarity with the uses of the New World balsam. Insisting that he has no direct knowledge of the production of its oil or its application, Oviedo maintains that he has had to rely on others for information: "In this I follow others who have experienced it, because I have not seen it used nor applied."[31] He offers differing opinions on the medical efficacy of the liquor, both positive and negative, albeit these too are based on others' reports: "I have heard great praise of the balsam . . . [and] also I have heard others curse it and say that it is dangerous when not applied well."[32] These protestations of ignorance are necessarily open to doubt. Oviedo's comments on the balsam's abundance on the islands and the mainland of the Indies suggest that many residents—including himself—were familiar with the plant.[33] He had obviously seen the tree before since he provides a detailed account of its appearance, texture, and smell.[34] So too, the diverse opinions he gathers regarding the benefits of its ointment indicate that it was widely used for its medicinal properties. After all, natural remedies purporting to heal open wounds—as the balsam's liquor was said to do—were highly sought after owing to the difficulties of life overseas. In the *Historia* (1535), Oviedo includes accounts of several other medicinal trees and plants that treat similar injuries, indicating their popularity among the colonists.

Indeed, Oviedo acknowledges that he knows a great deal more about the New World balsam than what he is willing to share. Initially, he defends his reticence as a cautionary measure, designed to protect his readers from unsound medical advice: "I could say more, according to what I have been told and even based on what I have seen of its effects, both negative and positive, but I don't want anyone to cure themselves by my words, nor do I desire such credit in medicine, because I didn't study it and it isn't my profession."[35] Yet the fact that the *Historia* (1535) is veritably teeming with such advice would seem to signal that the real reason for his secrecy lies elsewhere.

Oviedo reveals the true source of his discomfort momentarily, protesting, "But I don't want to understand any more about this water [from the plant] nor the balsam, as there are many here who, thanks to their experience, can speak more extensively about this topic than I can, and it is prohibited to make it."[36] Oviedo attributes his reluctance to engage thoroughly with the subject to the legislation issued to protect Villasante's monopoly over the plant and its healing liquor. Implied in these comments is the awareness that, were Oviedo to offer a more detailed account of the process of making the balsam ointment, he would risk implicating himself and his readers in the ban on such activities. His silence is a deliberate evasion: Oviedo cannot fulfill the demands of his task as a natural historian and disclose what he knows without infringing on the conditions of the agreement between Villasante and the Crown.

Reflecting further on the implications of Villasante's monopoly for the success of Spain's overseas venture, Oviedo warns of the dangers of making knowledge of the natural world a commodity. He writes, "And that one [Villasante] died without accomplishing what he promised. But I say what is public: regarding the treasure that was supposed to be had, it did not come to pass. Truthfully, if my opinion were taken into consideration, neither his Majesty would put such a prohibition on a thing that could result in so much good, nor would he stop ordering it to be made by as many as wanted, and after to distribute it among all those who had need for it; since the king cannot be lacking in other, greater interests for the increase of his rent."[37] Oviedo criticizes the protections issued to Villasante as having failed in their desired aim: the entrepreneur's death left his business languishing and the Crown no richer. The balsam's true value is not in the profit it generates, he argues, but in its benefits as a medicinal remedy. Putting the well-being of the Crown's subjects first and foremost, Oviedo demands that anyone be allowed to produce and distribute the balsam ointment. Be they remedies for pain, heat, or thirst, Oviedo maintains that God reveals nature's secrets to man for the good of all.[38] He protests the Crown's attempts to make the balsam an economically profitable commercial venture, advocating instead that knowledge about New World flora and fauna be freely available.

The legal protections afforded to Villasante have a direct impact on the kind and quality of information Oviedo divulges about the New World balsam, so much so that he is forced to compromise the values at the very heart of his natural history. Oviedo's insistence on *not* having seen and *not* knowing certain details of the preparation of the balsam's healing ointment are inconsistent with the importance he assigned to being an eyewitness to the phenomena about which he writes. Oviedo could very well have chosen not to dedicate an entry to the plant at all. That he does so is telling. The silences and negations in his account of the New World balsam are not only necessary protectionist measures but they also reproduce rhetorically the consequences of Villasante's monopoly for access to information about the natural world of the Indies. Effectively, this chapter becomes a platform from which Oviedo criticizes efforts to impose limits on the uses of New World nature.

Challenges to the New World Balsam in the *Historia* (1535)

In the *Historia* (1535), Oviedo describes several other plants that have similar medicinal properties as the New World balsam—namely, the artificial balsam (*bálsamo artificial*) and the *perebecenuc*. Unlike Oviedo's reticence with regard to Villasante's prized possession, however, his accounts of these other species are detailed, thorough, and based on first-person testimony. The differences between Oviedo's literary treatment of them and the New World balsam demonstrate just how far he deviates from his standard descriptive model to accommodate the restrictions that have been imposed on the latter.

Whereas Oviedo includes the entry on the New World balsam in *Book 10: On Medicinal Trees, and of Plants and their Properties*, the artificial balsam and the *perebecenuc* are not featured until later, in *Book 11: On Herbs and Seeds Brought from Spain to this Island, and of the Others that Here are Found and of Other Things*. Oviedo does indeed consider the *perebecenuc* an herb, so its placement in *Book 11* is fitting. But the same logic dictates that the artificial balsam would be better suited for *Book 10*: Oviedo categorizes it as a medicinal plant, one with the same ability to heal open wounds as Villasante's balsam. Although no explanation is given as to the decision to pair the artificial balsam with the *perebecenuc* in *Book 11*, in doing so Oviedo effectively separates the balsam from its peers. This narrative isolation foreshadows other rhetorical strategies that Oviedo employs to create distance between this healing plant and others like it.

At the same time as Oviedo places the New World balsam in a category by itself, somewhat counterintuitively he casts doubt on its singularity. Indeed, Villasante's balsam and the artificial balsam resemble one another to such an extent that occasionally Oviedo refers to the former as the latter.[39] It is as if Oviedo seeks to undermine the special commercial status of the New World balsam by blurring the boundaries between the two plants. Oviedo also notes that the balsam, artificial balsam, and *perebecenuc* are only a few of the many plants and herbs that exist to provide similar relief to humankind: "I believe that there must be other innumerable herbs and plants and trees suitable for our human sufferings and wounds."[40] The fact that Villasante's balsam is far from unique in its capabilities is an implicit retort to the privileged status its healing ointment has attained.

Furthermore, Oviedo openly calls into question the importance of the New World balsam by comparing the efficacy of its healing ointment with that of these other plants. He mentions several times that those who have used artificial balsam ointment find it "better and safer than Villasante's balsam or liquor."[41] Not content to offer such broad assurances, Oviedo recounts an anecdote that testifies to its power: a black man, at the point of death after flesh had been ripped away from his leg, used the ointment and recovered completely in less than twenty days.[42] Regarding the liquor of the *perebecenuc*, Oviedo reports that he often uses it to cure the flesh wounds of those natives and slaves who work in the fields.[43] These anecdotes are not unusual; after all, Oviedo's lived experience forms the premise of the *Historia* (1535). What is more striking is their absence from his account of Villasante's balsam. The general, contradictory judgments Oviedo claims to have heard concerning the New World balsam's curative abilities hold little weight when considered alongside the specific instances of healing that he relates regarding these other medicinal liquors.

Oviedo accompanies these success stories with lengthy and detailed instructions for the preparation of the artificial balsam's and the *perebecenuc*'s healing ointment. He provides first-person testimony of the parts and amounts of the plant to use and directions for how to boil the concoction and apply the resulting ointment to flesh wounds.[44] Of the similar process of manufacturing the liquor of the New World balsam, Oviedo has little to say, writing simply: "According to what some say, it is made of small pieces of this tree [the balsam], which are cooked and the water that comes out of them is a liquor."[45] His initial excuse for not being more forthcoming about what he knows in this particular case—reluctance to provide information that could be construed as medical advice—has been all but forgotten by *Book 11*, where Oviedo gives such instruction freely and abundantly.

A final feature of Oviedo's accounts of the artificial balsam and the *perebecenuc* absent from the description of Villasante's balsam is the use of visual material (Figures 27 and 28). Oviedo explains in detail the size and shape of the respective leaves of the former plants, as well as variations in their color and stems.[46] The accompanying illustrations put on display many of these characteristics. The woodcut of the artificial balsam leaf is particularly stunning, constituting one of the largest in the *Historia* (1535). Unlike other instances in which Oviedo laments his poor artistic skills, here he takes a certain pride in his drawing: "It seems to me that this leaf is not badly drawn but rather very well."[47]

The New World balsam's relative anonymity contrasts with the efforts Oviedo makes to bring the artificial balsam and the *perebecenuc* alive for readers. That he highlights no unique features of the plant via the inclusion of an image is perhaps of little note in and of itself—after all, most of the plants and animals chronicled in the *Historia* (1535) are not illustrated. However, given the other omissions that characterize Oviedo's account of the New World balsam, as well as his complaints about the negative impact of Villasante's monopoly over the plant, the absence of a woodcut is telling.

Indeed, Oviedo was well aware of the politic import of images, a fact clearly demonstrated in *Batallas y Quinquagenas* (1535–1557) when he is tasked with documenting the lives and genealogies of those who fought against the monarchy in the Revolt of the Comuneros (1520–1521).[48] Determined to punish them rhetorically for their treason, Oviedo announces his plans to condemn their actions. Through his mouthpiece, the Alcaide, Oviedo explains that he has eschewed his standard practice of devoting a stand-alone dialogue to each nobleman featured in the work, deeming it more appropriate to dedicate single paragraphs to the rebels. (He groups these, twenty in total, into a single dialogue that constitutes the first of the second *Batalla*.)

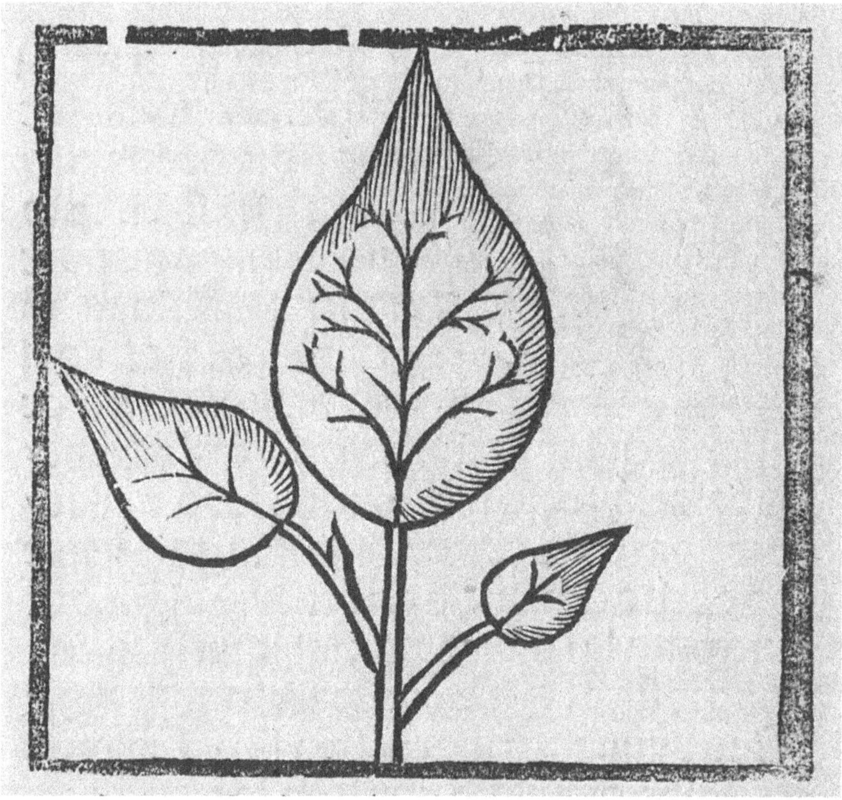

Figure 27. *Perebecenuc* leaf. *Historia* (1535), xcviii r [bk. 11, chap. 5]. Beinecke Rare Book and Manuscript Library, Yale University.

More importantly, Oviedo refuses to include illustrations of the rebels' crests (*timbres*), as he does in other instances.[49] Given the importance of historical memory for Oviedo, he is averse to omitting these individuals from *Batallas* entirely. Instead, he makes the most of both word and image to show that they are not as deserving of remembrance as others are.

Oviedo's rhetorical handling of the New World balsam in the *Historia* (1535) embodies a similar reaction to a situation with which he is fundamentally at odds. His denouncement of the legal measures taken to protect Villasante's claims to the balsam's healing liquor rings out no less true in what he leaves unspoken and unillustrated in the chapter. These silences become even more apparent when viewed alongside the chapters devoted to the artificial

Figure 28. Artificial balsam leaf. *Historia* (1535), cxvii v [bk. 11, chap. 4]. Beinecke Rare Book and Manuscript Library, Yale University.

balsam and the *perebecenuc* in *Book 11*. Oviedo's treatment of these plants—comparable to the New World balsam in their function as medicinal remedies—demonstrates the degree to which he deviates from his typical approach to the flora of the Indies when faced with Villasante's monopoly. Oviedo views these commercial arrangements as provoking an epistemic crisis, one that affects his task as a natural historian as well as the health and well-being of mankind. He enacts the consequences of this legislation through the tools of his trade, demonstrating the power of word and image to address not only the objectives of his natural history but also his grievances.

Crises of Knowledge: The Limits of Nature in the *Historia* (1535)

Oviedo is reticent about other flora and fauna of the Indies, too, albeit for reasons other than those at play in the case of Villasante's balsam. These instances attest to Oviedo's difficulties in articulating realities that defy his understanding. Many New World species of plants and animals do not have Old World equivalents, forcing Oviedo to describe their appearance, taste, smell, and behaviors via partial analogies with phenomena overseas. In bridging the gap between what is known and unknown, Oviedo strives to make familiar what is entirely foreign. There are times, however, when such comparisons are not forthcoming; the extreme and deceptive appearance and behavior of certain flora and fauna leave Oviedo at a loss to answer even the most basic queries about them. Above all, cacti and the iguana resist his efforts to describe and classify satisfactorily. Eluding easy placement in classical frameworks of knowledge, Oviedo's puzzlement about these species epitomizes the limits of European understanding of New World nature in the first half of the sixteenth century. Oviedo's response to such predicaments brings to the fore the importance he accords to illustration as a vehicle for making life at the very margins of known existence understood.

Freaks of Nature: New World Cacti and Questions of Identity

No plant family elicits such extreme emotions from Oviedo as that of the *Cactaceae*, of which the cactus is a member. These spiny, uninviting creatures mystify Oviedo to the extent that he despairs of making sense of them. His reaction is not surprising, considering that he had neither read about nor seen a plant like it; endemic to the Americas, cacti were complete novelties for Europeans. For this reason, none of the plants in Pliny the Elder's *Historia naturalis* belong to this family. The Roman historian does, however, relate Theophrastus's description of a plant whose root "throws out stalks which creep along the ground, the leaves being broad and thorny."[50] Pliny refers to this plant alternatively as a type of *carduus*—a thistle or cardoon—and *cactos*—a name derived from the Greek *kaktos* (cardoon). While the modern-day cactus takes its name from this classical counterpart, the two are not related: the thistles described by Pliny and Theophrastus were likely what we know today as the artichoke, a member of the sunflower family. Oviedo would have witnessed these and other kinds of thistles growing abundantly in Spain and Italy, where farmers considered them a nuisance.[51] His utter bewilderment upon encountering cacti suggests that he found few resemblances between thistles and these New World species of plants, besides their prickly nature. For want of any other European equivalent for these monstrous beings, however, Oviedo describes them generically as "a kind of thistle."[52]

Oviedo includes four species of cactus in the *Historia* (1535)—the pitahaya or dragon fruit, a barrel cactus (*cardos altos*), the prickly pear, and the welding tree (*árbol de las quebraduras*). He organizes their entries according to what he understands as the primary function of each plant: the first three cacti are fruit-bearing, and so comprise the final chapters of *Book 8: On Fruit Trees*. In contrast, the welding tree's healing properties prompt Oviedo to place it in *Book 10: On Medicinal Trees, and of Plants and their Properties*. Though distinct from one another, all four have the same fierce spines that repel even the slightest touch, and which perplex Oviedo. He is at a loss to describe their different parts as well, since they do not seem to correspond to the leaves and branches found on other plants. Effectively, cacti decrease or eliminate their leaves to reduce water evaporation, and the green stems—its branches—are bereft of bark and have a thick, waxy epidermis.[53] No wonder Oviedo is confused! He puts forth various hypotheses regarding their structure and growth, conjecturing, for example, that the pitahaya has "branches or long arms" instead of leaves.[54]

Oviedo's struggles extend to his ability to distinguish one species of cactus from another. Uncertain whether the prickly pear and the welding tree are truly unrelated species, Oviedo urges his readers to remember what he says in *Book 8* about the former when it comes time to discuss the latter in *Book 10*; the similarities between the two make him speculate on whether "these thistles convert into those trees."[55] At the same time, their differences give him pause: the prickly pear fruit is consumed, whereas that of the welding tree is made into a thick paste. Oviedo adds that the paste is so effective for painting that "on some of the things that I painted on paper the color is more vibrant now than it was six years ago."[56]

While Oviedo's inability to differentiate between the prickly pear and the welding tree might strike a modern-day reader of the *Historia* (1535) as preposterous, in truth the prickly pear (*Opuntia*) is notoriously difficult to classify due to its regional variety and ability to hybridize.[57] The features of the plant that Oviedo identifies—its flat, rounded stem segments that modern botanists call cladodes, as well as the groups of small, hairlike spines that dot its surface and are known as glochids—are the same markings used to identify the genus today.[58] Since the prickly pear takes the form of shrubs, trees, and creeping plants, it is not inconceivable that the plant Oviedo refers to as the welding tree is actually a type of prickly pear. Clearly, cacti upend all of Oviedo's expectations regarding how plants should look and behave. Never having encountered such beings before—either in his travels through Europe or in the works of classical and medieval authorities—Oviedo struggles to put into words a reality beyond his comprehension.

Unlike Oviedo's high praise of the pineapple, his interactions with the cactus highlight the anxieties and fears that formed the backdrop to many of the culinary discoveries the Spanish made overseas. Not coincidentally, Oviedo's first exposure to the prickly pear results in a situation where he believes his own death to be imminent. He recounts the circumstances as follows. Shortly after arriving in the Indies for the first time, in 1515, he disembarked on the island of Hispaniola along with several other companions. Veterans of the New World, Oviedo did not hesitate to follow their lead when he observed them picking the round, spiny fruit of the prickly pear and eating it. Several hours later, upon relieving himself, he observed with horror that his urine was red—pure blood! Instantly, Oviedo concluded that all the veins in his body had burst, and that he was literally watching his life drain away. Noticing the sudden pallor that had come over his companion's face, the pilot Andrés Niño approached the moribund man to ask what ailed him. "I am not in pain," Oviedo responded sadly, "but I would give up my horse and even another four to be close to Santo Domingo and to the great doctor *licenciado* Barreda, because without a doubt all the veins in my body have ruptured."[59] Much to Oviedo's surprise, his remarks were met with a great fit of laughter. No, his life was not in danger, the pilot assured him. It was the *tuna*, the fruit he had eaten, that had colored his urine red. Although retrospectively, in the *Historia* (1535), Oviedo refers to the fruit's effect as "amusing," this is certainly not how he must have felt when he imagined himself at death's door.[60] His initial encounter with the prickly pear brings to the fore the risks of life in the New World for Spanish colonists, the threat of harm or death present at every turn. That the prickly pear takes center stage in this anecdote is only fitting. A mysterious, frightful plant, the cactus confounds Oviedo's efforts to bring it into the sphere of the known. Representing the insuperable gap between seeing and understanding, the prickly pear is a metaphorical thorn in Oviedo's side.

Whereas the prickly pear causes Oviedo to come face-to-face with his mortality, the welding tree makes him despair at his inability to effectively communicate the nature of this being to his readers. Surprisingly, Oviedo's struggles in *Book 10* with the welding tree have little to do with the confusion he expresses in *Book 8* regarding the relationship between it and the prickly pear. Indeed, he barely seems to remember his earlier hypothesis that the two are one and the same; although he continues to note resemblances between their leaves, he makes this remark in passing and with the qualification that other flora, too, share this trait.[61] Apparently, Oviedo had resolved his initial doubts at some point prior to writing *Book 10*, concluding that their similarities do not imply that the two plants are the same species.

Instead, Oviedo's perplexity regarding the welding tree stems entirely from the enigmatic nature of the plant itself. He cannot establish with certainty even the most basic facts about it, beginning with the cactus's name. Surprisingly, he attributes his ignorance not to the plant's scarcity, but its ubiquity: "I do not know the name that the Indians of this island [Hispaniola] give to it nor its name on the other islands and the mainland because it is called something different in every region due to the multitude of languages that exist in the Indies."[62] Often when Oviedo encounters linguistic diversity, he provides various names for a given phenomenon, taking care to clarify the region where each is used. In this case, Oviedo has the opposite reaction: presented with a dizzying array of options, he claims to be unfamiliar with them all.[63] In lieu of an indigenous name for the plant, Oviedo decides to call it by its popular Spanish name—the welding tree, or *árbol de las quebraduras*, in reference to its healing properties.

Its name is only one of Oviedo's difficulties in dealing with this intractable botanical beast. An even greater problem for him is determining the welding tree's manner of growth so as to make sense of its disparate parts. Unable to discern the plant's branches from its leaves, ultimately Oviedo concludes that any difference between them is negligible because one morphs into the other: "These branches were first leaves, which hardened and turned into branches. And these hardened leaves, or while they are in the process of hardening, procreate others, and the leaves turn into branches."[64]

This "monster among trees" proves impossible for Oviedo to describe to his satisfaction.[65] He turns to skilled Renaissance painters, appealing to them to bring the plant to life: "It is better to see it painted by the hand of Berruguete or another excellent painter like him, or Leonardo da Vinci or Andrea Mantegna, famous painters that I met in Italy, than to attempt to explain it with words."[66] Having no other choice, eventually Oviedo assumes the task himself (Figure 29). He expresses just as much disappointment in the final product of his artistic efforts as he does in the ability of the written word to communicate the true nature of the species: "As best I can, I will put here the form that this tree has with my badly executed drawing, so that along with what I have said it is easier to understand and consider."[67]

The woodcut of the welding tree takes its place alongside those of the prickly pear, the barrel cactus, and the pitahaya (Figures 31, 32, and 30). Oviedo deviates from his standard approach to illustrating flora in the *Historia* (1535) in all but the woodcut of the latter. Instead of representing these species visually via their fruit or leaves, as he does in other instances, Oviedo portrays them as complete specimens. His inability to make rhyme or reason of cacti likely accounts for this artistic decision; Oviedo is no more able to identify and isolate one part of the plant from another visually than he is in writing. His dependence on illustration to communicate what he is at pains to understand highlights the importance of this medium for Oviedo when faced with nature's challenges.

Cacti exert a powerful hold over Oviedo, simultaneously mystifying, frustrating, and fascinating him. The prickly pear and the welding tree expose him to a natural world for which no number of partial comparisons with overseas phenomena can do justice. Having exhausted the potential of word and image to make this reality understood, Oviedo offers one final option to those readers interested in learning more about the welding tree. He invites them to examine the plant in situ, going so far as to provide directions to a locale on the island of Hispaniola where they grow.[68] Oviedo must have been aware that to make the journey to the Indies, much less search out the welding tree, would have been an impossible aspiration for most of his readers. Yet these strange beings offer him few alternatives. Teetering precariously at

Figure 29. Welding Tree. *Historia* (1535), xcii r [bk. 10, chap. 1]. Beinecke Rare Book and Manuscript Library, Yale University.

the edge of the known universe, Oviedo is all too aware of the limitations of his knowledge when faced with life in the New World.

Flesh or Fish?: The Curious Case of the Iguana

Certain New World fauna prove just as difficult to incorporate into early modern European categories of knowledge as cacti. Indeed, even the most everyday species present challenges to Oviedo in his efforts to classify them. The iguana falls into this group: beloved by Spaniards and indigenous people alike as a cuisine, nevertheless its identity remains elusive.[69] It is so frightful

Figure 30. Pitahaya. *Historia* (1535), lxxxv v [bk. 8, chap. 23]. Beinecke Rare Book and Manuscript Library, Yale University.

to behold that Oviedo maintains it would take great courage to stand one's ground upon encountering it for the first time, let alone to try its flesh.[70] Its sharp teeth, the spiny protuberances that line its backbone, and its birdlike claws are at odds with its docility and pleasant taste. Compounding Oviedo's difficulties in describing the iguana is its enigmatic behavior: the lizard seems equally at home in the water as on land, leaving him uncertain whether it is animal or fish. He also disputes the royal chronicler Pietro Martire d'Anghiera's prior attempts to classify the iguana, which to Oviedo's mind only serve to further confuse the issue. The Spanish historian's account of the iguana in the *Historia* (1535) represents his most vigorous defense of the authority of the eyewitness to New World wonders, as well as his most acute struggle to resolve problems unanticipated by his classical predecessors.

Figure 31. Prickly Pear. *Historia* (1535), lxxxvi v [bk. 8, chap. 25]. Beinecke Rare Book and Manuscript Library, Yale University.

It is common knowledge that, for Oviedo, Pietro Martire d'Anghiera's *De orbe novo decades* (*Decades of the New World*) (1516, 1530) epitomizes the hazards of armchair historians—that is, those who acquire knowledge of the Indies in relative comfort, without undertaking the difficult overseas journey. On several occasions in the *Historia* (1535), Oviedo claims to identify errors committed by the Italian humanist, ranging from cases of mistaken identity to islands and trees that simply do not exist.[71] These remarks are often made summarily, tangential to Oviedo's endeavors to document a given phenomenon. In contrast, Oviedo treats the inaccuracies he finds in Martire's description of the iguana much more severely. The crux of his disagreement with the Italian humanist is that "Pietro Martire . . . says that these iguanas

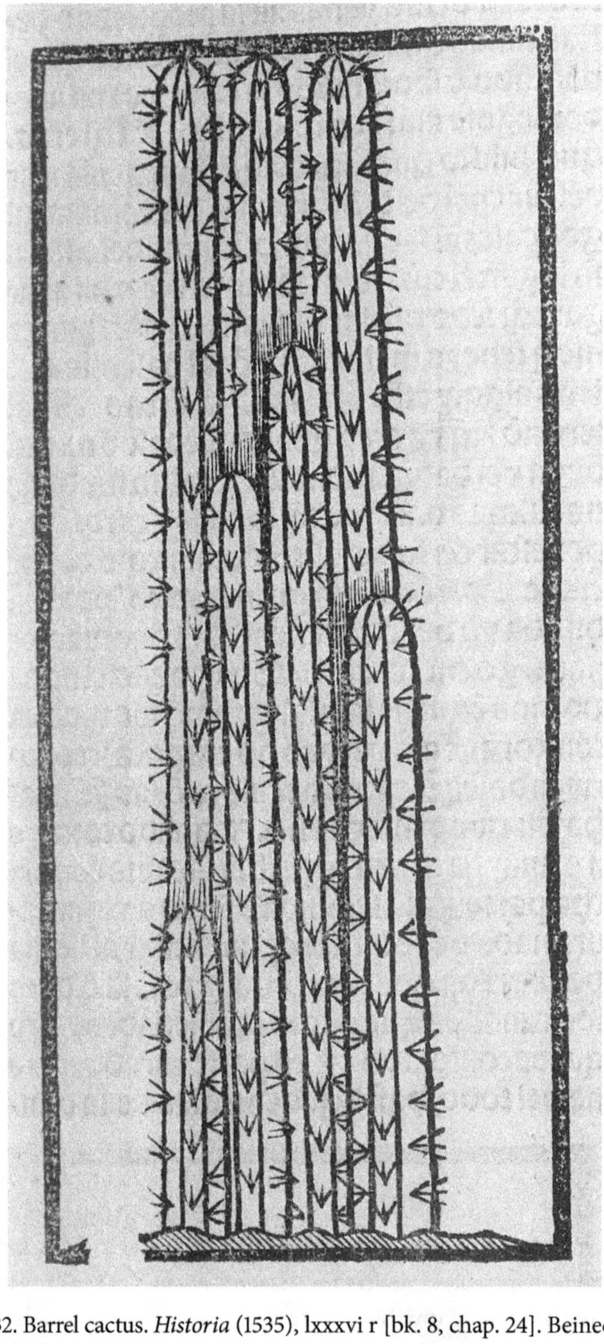

Figure 32. Barrel cactus. *Historia* (1535), lxxxvi r [bk. 8, chap. 24]. Beinecke Rare Book and Manuscript Library, Yale University.

are similar to the crocodile, which are animals from the Nile River."[72] Oviedo dedicates roughly half of his chapter on the iguana to refuting this idea. Employing Pliny the Elder's (23–79 CE) and Saint Isidore of Seville's (d. 636 CE) descriptions of the crocodile, Oviedo seeks to demonstrate the innumerable differences between this creature (which he had never seen) and the iguana. These include the former's large size, its color, its lack of a tongue, and its ability to move only its upper jaw.

Oviedo excuses his lengthy comparison of the two animals as an effort to correct the errors in Martire's account: "and if I have dragged this out so much, it has been to banish Pietro Martire's opinion from readers."[73] In refusing to give any credit to the Italian humanist's account of the iguana, however, Oviedo overlooks the nuances in Martire's argument that provide support for Oviedo's own opinions. Although Martire notes certain resemblances between the iguana and the crocodile, he also observes a key difference: "[Columbus and his men] affirm that [the iguana] does not differ in the least bit from the Egyptian crocodile, except in their size, since Pliny says that some crocodiles were found that were eighteen cubits but the largest of these serpents are eight feet."[74] In the *Historia* (1535), Oviedo notes a similar disparity, maintaining that, on average, the iguana measures 4 feet (from the tip of the tail to the top of the head) and that the largest specimen he has seen measured a little less than 5 feet.[75] He cites testimony from Pliny and Isidore that the crocodile is 18 cubits long (29 ½ feet) and 20 cubits (30 feet), respectively.[76]

Even though Martire's account highlights the very point Oviedo seeks to make—that the iguana is significantly smaller than the crocodile—he refuses to acknowledge their shared viewpoint. Quite to the contrary, he insists on Martire's ignorance of the subject, writing, "[Martire] is much-deceived, and those who write of these things from hearsay are forced to commit similar and notorious errors because these iguanas are not any larger than what I have said."[77] Disregarding any common ground in their understanding of the iguana, instead Oviedo takes advantage of the opportunity to demonstrate the dangers of writing without firsthand experience. Were Martire to have seen these iguanas for himself, Oviedo complains, he would not have related such baseless assertions.

So insistent is Oviedo on the differences between the iguana and the crocodile that he ignores the parallels he indirectly observes between the two species. Oviedo includes references from Isidore and Pliny as to the semiaquatic nature of the crocodile; in a separate passage, Oviedo mentions that the iguana resides both in the water and on land. Nowhere, however, does

Oviedo acknowledge this to be a shared behavior. Perhaps he dismisses this characteristic as constituting grounds for comparison, or fears readers will misinterpret such attempts. Whatever the case may be, Oviedo's unwillingness to draw connections between the iguana and the crocodile is unusual for a natural historian who relies on such parallels between species in the Old World and the New to make himself understood. Oviedo discounts any possibility of apprehending this creature through a comparative framework; the iguana stands alone as a novelty of the New World.

Ironically, it is the iguana's behavior as a creature of both the land and water that presents the classification difficulties Oviedo confronts time and time again in his endeavors to describe it.[78] Initially, in the *Sumario* (1526), he resolves the debate as to whether the iguana is animal or fish by categorizing it among the "vipers, snakes, serpents, lizards, toads, and other, similar things."[79] In the *Historia* (1535), Oviedo reconsiders this placement, ultimately determining that the iguana is an aquatic creature. He is not entirely convinced by this decision, however. Expressing his doubts, Oviedo argues that the identity of the lizard is yet unknown; although he includes its entry in *Book 13: On Animals of the Water*, he defines the creature as a "neutral animal" because "Christians do not know how to determine if it is a type of flesh or fish."[80]

At first glance, it is difficult to make sense of Oviedo's confusion regarding the iguana's aquatic and terrestrial habits. Native to the New World, the reptile does not appear in Pliny's *Historia naturalis*. Even so, the Roman historian describes several animals that behave similarly to the iguana: in addition to the crocodile, Pliny also describes a bramble frog and a sea calf that live equally in water and on land.[81] And while Isidore had already classified creatures that evinced this kind of behavior as amphibians, it appears that Oviedo still considered the iguana a kind of anomaly.[82]

The root of Oviedo's perplexity and preoccupation with the lizard can be traced to its culinary importance for Spanish colonists in the New World.[83] Had it not proven such a delectable treat, chances are that Oviedo would not have continued to wrestle with the question of its classification for so long. The popularity of its meat brings to the fore a religious dilemma unanticipated by Oviedo's classical predecessors—namely, should it be treated as flesh or fish on days of abstinence, as decreed by the church? It is not until the Monserrat manuscript, written between 1535 and 1549, that Oviedo explicitly addresses this issue. He revisits his former classification of the iguana as fish from the *Historia* (1535) and finds it wanting.[84] Instead, he decides to include the creature among terrestrial animals—as Pliny had done with the crocodile—accompanied by the categorical statement: "I consider it meat. I do not say this so that people stop following their own conscience, and principally what is ordered by prelates and the church."[85] Oviedo's decades-long struggle in determining the true nature of the iguana is not a question of scientific interest but rather of spiritual concern. His attempts to resolve these queries embody the challenges that New World phenomena pose to different facets of colonial life.

Not coincidentally, Oviedo makes the decision to include an illustration of the lizard in a moment in which he has yet to find an answer to this dilemma (Figure 33). As with all the visual material in the *Historia* (1535), he hopes the woodcut will allow readers to better understand the creature: "This that I painted here wants to resemble the iguana, and truthfully this is the form they have."[86] Nevertheless, Oviedo is acutely aware of the limitations of such pictorial efforts, and he urges caution when viewing it: "with the protestation or warning that I mentioned at the beginning of this chapter, that this animal is neutral and men cannot determine whether it is of the land or the water."[87] The act of seeing the iguana, be it in person or via an illustration, does not shed any additional light on the question of its identity. In other words, this woodcut embodies the same barriers to knowledge present in Oviedo's written account of the iguana. As much as Oviedo avails himself of the visual medium to help decipher New World nature, neither word nor image provides responses to his queries about the iguana.

Of all the flora and fauna that Oviedo describes and illustrates in the *Sumario* and the *Historia general*, none poses more of a challenge than the iguana. Oviedo's struggle to represent and classify this creature continues for his entire career, reflecting his confusion over its appearance as well as its behavior. More than any other New World fauna in the *Historia* (1535), the iguana symbolizes that which is known and widely celebrated, but also unknown and mysterious. In Oviedo's attempts to describe this being, he is taken to the limits of European knowledge of the Americas. In this fraught situation, even illustration cannot aid him in making sense of the iguana's utter novelty.

Figure 33. Iguana. *Historia* (1535), ciii v [bk. 13, chap. 3]. Beinecke Rare Book and Manuscript Library, Yale University.

Balsams, Cacti, and Iguanas: Conclusions

While the challenges of Oviedo's endeavors to describe strange beings like cacti and the iguana are different from the struggles that occupy him in the case of the New World balsam, similar concerns underlie all three. Be it in response to Villasante's monopoly or the enigmatic nature of certain New World phenomena, Oviedo's encounters with these species prompt epistemological crises that expose the shaky foundations on which his natural history project rests. Unable or unwilling to communicate in words what these flora and fauna are truly like, Oviedo alternatively shares and conceals, admits confusion and plays at ignorance. In these pivotal moments, Oviedo wields his artistic impulses strategically, as an ultimate recourse to both facilitate and deny understanding. At the precipice of the unknown, Oviedo draws his boundaries carefully, ever cautious not to step over the edge.

CHAPTER 7

A World United

The Monserrat Manuscript's Natural History

Even before the publication of the *Sumario* in 1526, Oviedo had already conceived a much more ambitious project.[1] The final product of this undertaking, which remained a work in progress from the end of the 1530s through 1549, was the monumental *Historia general y natural de las Indias*. Consisting ultimately of three parts divided into four volumes, the work's sweeping narrative transcends the limited geographies of its predecessors. Whereas the *Sumario* and the first part of the *Historia* (1535) revolve around life in the Caribbean, the three-part *Historia general* embraces the human and natural histories of territories as far north as Florida and as far south as what today are Peru, Chile, and Argentina. The ultimate testament to Oviedo's lifelong efforts to document the far-flung realities of the Indies, this accomplishment would not be fully appreciated until long after his death. Unable to finance its printing, in 1549 Oviedo entrusted the manuscript of the *Historia general* to an unidentified monastery in Seville.[2] Although attempts were made to publish the work in the eighteenth century, it was not until 1851 that the Real Academia de la Historia issued the first of three volumes (1851–1855) of the *Historia general* under the editorial supervision of the Spanish historian José Amador de los Ríos (1816–1878).[3] In the 1940s, the work was published again (ed. Juan Natalicio González, 1897–1966), and for a final time in 1959 in the series Biblioteca de autores españoles (ed. Juan Pérez de Tudela Bueso, 1922–2004).

Although these historians merit recognition for having drawn attention to Oviedo's role as an early chronicler of the Indies, their editorial interventions complicate our understanding of the *Historia general* as it developed over time.[4] Particularly problematic are the changes they made to the Monserrat manuscript—that is, the parts of the *Historia general* conserved in Oviedo's hand. (The Monserrat manuscript is denominated as such because it was donated to the Monserrat monastery by the widow of don Luis Salazar y Castro upon her husband's death in 1734.)[5] Above all, Ríos—to whom both Natalicio González and Tudela Bueso defer in their editorial judgments—imposed a uniformity on the work that is noticeably absent from the original.[6] Far from the monolithic, homogeneous text of modern editions, extant portions of the Monserrat manuscript reveal that Oviedo continued to revise the *Historia general* extensively for fourteen years (1535–1549).[7] The edits that Oviedo made to the manuscript attest to the ways in which he accommodated new information and dealt with the challenges of chronicling the human and natural history of a constantly expanding empire.

Scholars, especially Jesús Carrillo Castillo, have shed light on Oviedo's working methods in the manuscript.[8] He argues that Oviedo's editorial interventions attest to several distinct phases in its composition. For Carrillo Castillo, the defining moment in the project occurred roughly six years after the work's inception, between 1541 and 1542. At this point, Carrillo Castillo posits, Oviedo made profound structural changes to the *Historia general*'s natural history, which impacted the entire manuscript. Initially, Oviedo had organized this material according to geographic criteria: each of the three parts of the work was imagined as containing an account of the natural and human history of a particular region—the Caribbean, in the case of part 1, and the Atlantic and Pacific coasts, in parts 2 and 3, respectively.[9] In 1541, however, Carrillo Castillo argues that Oviedo determined to abandon this tripartite layout in favor of consolidating the work's natural history into part 1.[10] In essence, and thanks to these changes, what had once been a description of Caribbean flora and fauna effectively became one of New World nature at large.

In this chapter, I propose an alternative chronology for these revisions. I maintain that Oviedo decided to restructure the *Historia general*'s natural history not in 1541 but rather several years earlier, upon embarking on the manuscript. Although this move had profound implications for the work as we know it today, it did not entail major shifts in Oviedo's literary treatment of nature. Indeed, this decision was a response to considerations similar to those that had informed the organization of the earlier *Sumario* and *Historia* (1535). In this sense, the changes to the Monserrat manuscript are consistent with Oviedo's previous attempts to address the descriptive and structural dilemmas posed by the plants and animals of the Indies. His handling of these topics in the Monserrat manuscript is at once the culmination of decades of inquiry, observation, and writing about New World nature as well as their ultimate expression.

The Monserrat Manuscript's *Book 12*

Instrumental to the argument laid out here regarding the Monserrat manuscript's natural history is my access to the manuscript of *Book 12: On Animals*, long since presumed lost. Modern scholarship has relied almost exclusively on the four extant books dedicated to New World nature (of ten in total) whose original manuscripts are in the Huntington Library: *Book 6: Of Deposits*, *Book 7: On Agriculture*, *Book 9: On Wild Trees*, and *Book 11: On Herbs and Seeds*. The disappearance of the remaining six books, including *Book 12*, was the result of losses the manuscript suffered sometime after 1549. The Spanish historian Juan Bautista Muñoz (1745–1799) first recorded the absence of four of the six books—books 10, 13, 14, and 15 (*On Medicinal Trees and the Plants and their Properties*, *On Animals of the Water*, *On Birds*, and *On Insects*, respectively)—in the second half of the eighteenth century.[11] Although these books appeared initially in the *Historia* (1535), Oviedo continued to add to them after the publication of the first edition: the Monserrat manuscript's table of contents indicates that Oviedo included seven more chapters in *Book 10*, five in *Book 13*, twenty-four in *Book 14*, and sixteen in *Book 15*.[12]

Still more of the Monserrat manuscript's natural history books went missing or changed hands in unknown circumstances after Ríos completed his edition of the *Historia general* (1851–1855), including the still-lost *Book 8: On Fruit Trees*.[13] Around World War I, Dr. Maurice Ettinghausen purchased ten books of the Monserrat manuscript from an unidentified seller on behalf of the antiquarian bookdealers the Maggs Brothers (est. 1853); a number of these were dedicated to New World nature—namely, books 6, 7, 9, 11, and 12. On February 25, 1926, the Huntington Library purchased seven of these (including 6, 7, 9, and 11), where they remain today. In the meantime, however, the books the library did not purchase—12, 17, and 19, as well as a table of contents of the first nineteen books, complete with chapter headings—vanished.[14] Yet they continued in the Maggs Brothers' possession until 2016, when they were sold to an anonymous buyer.

The manuscript of *Book 12* is integral to understanding Oviedo and his natural history project. Despite the untold merits of the books at the

Huntington Library, they provide an incomplete picture of New World nature in the *Historia general. Book 7: On Agriculture*, *Book 9: On Wild Trees*, and *Book 11: On Herbs* are dedicated exclusively to American flora, whereas only a handful of chapters in *Book 6: Of Deposits* describe species of animals, fish, or birds. *Book 12: On Animals* is a necessary complement to these books, containing the only comprehensive account of New World fauna in the work, as well as the only hand-drawn illustrations of them.

Book 12's manuscript is important too for the change it registers from its earlier appearance in the *Historia* (1535). Having only five chapters in the latter, originally *Book 12* was by far the shortest natural history book and one of the few that did not include any woodcuts. By the time Oviedo completed the manuscript, however, it had grown by thirty-two chapters, to a total of forty—five times its size in the *Historia* (1535). Such an increase dwarfs the expansions of books 7, 9, and 11, which grew by six, eighteen, and seven chapters, respectively.[15] So too, whereas *Book 12* originally included no woodcuts, the Monserrat manuscript features ten drawings (including one of the iguana, which initially had been in *Book 13*, as well as two of the anteater, one of which is crossed out). Again, this number of new drawings exceeds those added to other books: two to *Book 7*, three to *Book 9*, and four to *Book 11*. For these reasons, *Book 12: On Animals* provides us with a more nuanced awareness of the editorial decisions Oviedo made with respect to New World nature in the years following the publication of the *Historia* (1535).

New World Nature in the Monserrat Manuscript

While Carrillo Castillo traces the development of the Monserrat manuscript from its beginnings in the mid to late 1530s to 1549, here I concentrate on the initial years that Oviedo was at work on the project, from roughly 1535 to 1542, as most crucial to the work's natural history.[16] I construct my argument on the basis of the Monserrat manuscript's extant natural history books—namely, on books 7, 9, 11, and 12.[17] Inarguably, Oviedo's decision to consolidate the *Historia general*'s natural history into part 1 was of greater consequence for the work's final form than any other. Yet there is no consensus as to when this reorganization occurred. Carrillo Castillo dates these changes to 1541, in a moment of sudden and radical reshaping of the place of New World nature in the manuscript. Conversely, I maintain that Oviedo made this decision almost as soon as he set pen to paper to write his magnum opus, in the final years of the 1530s. Indeed, the argument can be made that Oviedo's revisions to the Monserrat manuscript were decades in the making, the result of his continued struggle to impose order on a natural world that was both one and many, united and divided.

The crux of Carrillo Castillo's argument lies in the differences that he purports exist between the base copy of the manuscript, produced sometime between 1535 and 1541, and its subsequent revisions (1541–1542). According to him, this core text consists of a "revision of the first part and a first version of the second [part]" of the *Historia general* that Oviedo intended to publish in the early 1540s.[18] From the original dates that Oviedo penned in books 4, 5, 6, and 8 (which are obscured to varying degrees, given that he constantly updated them), Carrillo Castillo concludes not only that the manuscript was written at the end of the 1530s but also that it continued to adhere to the geographical coordinates of the *Historia* (1535)—that is, the Caribbean. It was only when these publication plans failed, sometime in 1541, that, Carrillo Castillo argues, Oviedo determined to shift part 2's and part 3's natural histories into part 1, transforming the latter into a catchall for New World nature.

Even if it were possible to establish a relationship between the dates of composition and the geographical coordinates of the base copy of the Monserrat manuscript, however, such a wide-reaching argument is irrelevant. The issue at hand is not whether part 1 continued to revolve around life in the Caribbean in the early stages of its composition—very definitely, and as Oviedo intended, its human history (books 1–5 and 16–19) *does* remain anchored in this space. Rather, the question to be addressed concerns the contours of its natural history (books 6–15). Of the books Carrillo Castillo analyzes, books 4 and 5 are of little help in this inquiry; they recount aspects of the Spanish governance and administration of Hispaniola and, in the case of *Book 5*, indigenous customs on the island. Regarding those that pertain to natural history proper, only *Book 6* exists in Oviedo's hand; Carrillo Castillo is unable to date the passages he cites from Muñoz's transcription of *Book 8*.[19] Given these limitations, no conclusions can be reached from the dates that Carrillo Castillo provides regarding the geography of part 1's natural history in these early years.

Nor does Oviedo give any indication that the work he hoped to publish in the early 1540s respected the original structure of the *Historia* (1535). Carrillo Castillo interprets an undated passage from *Book 8* of the Monserrat manuscript as evidence that the failure to print this revised edition was the catalyst for the transformation of part 1's natural history: "Now that the time has arrived for the reprinting of this expanded and revised first part (and also the second and third), it has seemed to me that, so that the materials go together, they are placed in a way so that the reader does not have to go searching for my promises."[20] Yet the fact that this quote is taken from an undated transcription of the original manuscript means that alternative readings are possible. It is conceivable, for example, that Oviedo penned this sentence sometime toward the end of the 1530s, and that he added the parenthetical reference to parts 2 and 3 once his publication plans changed. Recall that neither Juan Bautista Muñoz nor José Amador de los Ríos—the only two to have worked with the manuscript of *Book 8* before it went missing in the early twentieth century—distinguish between the base copy of the text and small edits that Oviedo made to it over time. It is plausible that Muñoz and Ríos could have copied this passage in its entirety without denoting any parts as later additions. If this were the case, the quote in its original form would likely indicate that Oviedo had decided to make part 1 the repository for the work's natural history contemporaneous with his plans to publish the "expanded and revised first part."[21]

Such an interpretation is consistent with other edits Oviedo made to the *Historia* (1535) in the base copy of the Monserrat manuscript. Passages in the prologues of books 9, 11, and 12 in the print edition of the first part of the work speak to Oviedo's initial plans to include entries on flora and fauna in the still-to-be-written parts 2 and 3 of the work. In *Book 9: On Wild Trees* of the *Historia* (1535), Oviedo assures his readers that he will address other species of wild trees later, in relation to mainland territories: "There is much more to note there in all of these things, and in each one of them and others that are reserved for [these other parts], because they pertain to these lands and not to this one."[22] Oviedo makes similar remarks in the prologues to *Book 11: On Seeds and Herbs* and *Book 12: On Animals* in the *Historia* (1535). In the former, he alerts his readers that "in the second part of this natural and general history of the Indies (and in the third) will be found many other particularities and new plants and herbs that I set aside for their proper places and provinces."[23] Likewise, Oviedo excuses *Book 12*'s brevity with assurances that, "in the second and third parts, when the things on the mainland are addressed, there will be much more to write about this material because there are many and very different animals from those in our Spain."[24] There is no question that, prior to 1535, Oviedo envisioned New World nature as distributed among the work's three parts, in accordance with its geographical division.

The modifications Oviedo makes to these passages in the base text of the Monserrat manuscript suggest that he reconceived of this format soon after beginning the post-1535 project. In the prologue to *Book 9*, Oviedo explains that he no longer finds it prudent to withhold further information about wild trees until the second and third parts of the *Historia general*. Instead, he continues, "So as not to delay, and so the experience is moderate and what is known about the Indies currently is not put off too long, it appears to me that it will be best to include here in this book what is related to wild trees . . . so that the material is together and not scattered nor divided by the second and third parts of these books."[25] Oviedo makes similar revisions to the prologues of *Book 11* and *Book 12*: in the case of the former, he makes its geography more inclusive, moving away from a narrow focus on Hispaniola to embrace "other islands and the mainland," and eliminating any reference to his previous plans to address certain plants and medicinal herbs in parts two and three of the work.[26] Regarding *Book 12*, Oviedo explains in the prologue that it will include entries on the animals of Hispaniola followed by others devoted to fauna at large in the Indies: "Having done this [that is, spoken of the animals on Hispaniola] I will address all the other animals (which were not spoken of in the first edition of this first part of the *General History of the Indies*) . . . in whatever islands or parts of the mainland, ascribing to each animal its homeland . . . so that this type of material is all together."[27]

Collectively, these revisions indicate that Oviedo's priorities with regard to New World nature changed drastically between the *Historia* (1535) and the writing of the base copy of the Monserrat manuscript. In the former, Oviedo's overriding concern is to situate phenomena in the natural environment to which they belong; in the latter, thematic coherence takes precedence over geography. To be sure, Oviedo continues to call attention to the geographical distribution of species, although this criterion does not play such a central role in the overall organization of New World nature as before. Nature is no longer fragmented; in the Monserrat manuscript, Oviedo holds it out to readers as one united whole.

Oviedo's early decision to consolidate his account of natural history into part 1 of the *Historia general* explains the lack of extensive revisions to the base copy of the manuscript. Had he completed a revised version of part 1 and a preliminary version of part 2 prior to restructuring the work, as Carrillo

Castillo suggests, one could imagine that these alterations would have left profound traces in the text. The scholar lends weight to his argument by pointing to one such update Oviedo makes: he modifies a reference in *Book* 9 to an account of the cacao tree as being in a future book "up ahead"—ostensibly in part 2—with its new location in "the previous book [8]."[28] Other comparable corrections, however, are virtually absent from the manuscript, casting doubt on the probability that such drastic changes were made at the late date of 1541.[29]

A final line of reasoning that Carrillo Castillo advances to support his claims concerns the marginal notes found in part 1's books on flora and fauna. He posits that Oviedo moved them there from their original locations in parts 2 and 3 of the Monserrat manuscript, respectively, once he had decided to reorganize the work's natural history.[30] Such an argument takes for granted that Oviedo mentions no mainland species in the base copy of part 1 and, therefore, that the information he added in the margins indicates a profound, and late, shift from a focus on the Caribbean to the Americas at large. However, *tierra firme* appears frequently in the base copy of books 6, 7, 9, and 12. Some of these references are carryovers from the *Historia* (1535), but others are new additions that Oviedo included about the mainland counterparts of island species.[31]

Often these revisions are superficial but nevertheless telling, such as the numerous changes Oviedo makes to chapter titles from the *Historia* (1535) to encompass a wider geography.[32] In other instances, his interventions are more substantial; for example, Oviedo updates the entry on the ceiba and a breed of small dog (*perro gozque*) to reflect his experiences on the mainland.[33] His account of the palm undergoes noticeable changes as well.[34] Regarding the latter, Carrillo Castillo asserts that a marginal note in the chapter describing two species of mainland palm, the *pixabax* and *cañaspalmas*, confirms the natural world of the *tierra firme* as a late addition to part 1. Yet Oviedo had already incorporated three new species of mainland palm into the base copy of the chapter, including an extensive account of the coconut. Far from indicating the drastic transformation of part 1's natural history, then, the marginal notes on the *pixabax* and *cañaspalmas* are simply a consequence of Oviedo's ongoing interactions with New World nature. The overwhelming presence of mainland flora and fauna in the core text of part 1 of the Monserrat manuscript leads us to the conclusion that Oviedo made a concerted effort to expand the scope of part 1's natural history at the very outset of his post-1535 project.

New Worlds and Old Dilemmas

The conditions of Oviedo's appointment as chronicler of the Indies (*cronista de Indias*) were likely the immediate impetus prompting him to rethink his approach to New World nature in the *Historia general*. The roots of this decision go much deeper, however, and can be traced to his earlier difficulties in the *Sumario* and the *Historia* (1535) in giving expression to New World nature. Oviedo's decision to make part 1 of the Monserrat manuscript a catch-all for the flora and fauna of the Indies at large was neither unexpected nor innovative; rather, it responds to the same editorial concerns that characterize his previous writings on the Indies.

Six years after the publication of the *Sumario* and several years prior to that of the *Historia* (1535), Oviedo was awarded the position of chronicler of the Indies. Regarding the terms of the appointment, the Council of the Indies (*Consejo de Indias*) petitioned the Crown on May 27, 1532, that Oviedo be tasked with "traveling to those lands where he has not been, to see what he has not seen."[35] This reference to travel disappears in the royal decree issued three months later, on August 18, 1532; Oviedo is ordered simply to write about the Indies "thoroughly and with good style."[36] Regardless of whether Oviedo negotiated these new terms, as Carrillo Castillo argues, they drastically reimagined his role as chronicler.[37] Oviedo is no longer the roving observer described in the Council of Indies' petition; rather, he is a writer and compiler.[38] These modifications to his historiographical duties were of the utmost consequence for the writing of the *Historia general*. Above all, they impacted Oviedo's ability to acquire information about regions he had never visited, such as those known today as Mexico and Peru. To be sure, Oviedo was able to draw on others' accounts of these lands, but often these provided few insights into New World nature, focusing instead on the events of discovery, conquest, and colonization.[39] Given such limitations, it is likely that Oviedo determined that it was no longer feasible to include a stand-alone natural history in each of the three parts of the *Historia general*.

Indeed, the Monserrat manuscript attests to Oviedo's difficulties in acquiring information on the natural world farther afield than his own lived experiences. Although Oviedo continuously updates the *Historia general* with the most recent Spanish activities in the New World, for the most part the growth of its natural history is unrelated with the new geographies Oviedo chronicles. A perusal of the entries Oviedo adds to books 7, 9, 11, and 12 reveals that most deal with species from regions with which he was already well acquainted.[40] One of the most striking examples of this tendency is the manuscript of *Book 12*. Although it undergoes one of the largest expansions of any book devoted to New World nature—from eight chapters in the *Historia* (1535) to forty in the Monserrat manuscript—a relatively small number of these chapters are "new" in the strict sense of the word. Sixteen of the chapters Oviedo adds to the manuscript are devoted to mainland fauna from the *Sumario* that he had excluded from the *Historia* (1535) because of its narrower geographical focus. Only the last ten chapters in the Monserrat manuscript's *Book 12* feature animals that were not included in Oviedo's previous printed works. They span a vast geography, from llamas in New Castile to the *ayre* in Cuba and the buffalo in North America, in territories often distant from Oviedo's home on Hispaniola and tangential to his lived experiences in the Indies.[41]

These new entries share several common characteristics: in general, they are brief since Oviedo is forced to rely on other eyewitnesses to supply him with information or, alternatively, with specimens that have been removed from their native habitat. Oviedo relates that he learned of the bears in the New Kingdom of Granada from "the captains Juan de Junco y Gomez de Corral, and from others," and he makes similar statements with regard to an animal he refers to as an *aserrador* as well as the "cows," or buffalo, in North America.[42] Oviedo acknowledges having seen a llama, although it was a specimen brought to Hispaniola from New Castile, making his knowledge of the species necessarily limited.[43] These difficulties are reminiscent of those that Oviedo faces in the *Historia* (1535) in acquiring information on fauna on Hispaniola that were virtually wiped out by the first Spanish colonizers. In both instances, the information Oviedo is provided with lacks the thoroughness that defines much of his first-person testimony.

The manuscripts of books 7, 9 and, 11 similarly contain few new chapters devoted to species that Oviedo did not know firsthand. The six entries that Oviedo adds to *Book 7* all concern plants readily found on Hispaniola.[44] By the same token, only two of the fifteen chapters added to *Book 9* describe trees Oviedo had not encountered personally, a so-called cinnamon tree in Quito (chapter 31) and a willow tree in New Castile (chapter 32).[45] (Oviedo describes the former based on pods his friends and acquaintances sent to him in Santo Domingo.) The remainder (a total of thirteen chapters) provides accounts of species in regions with which he was well acquainted—namely, Hispaniola, Castilla del Oro, Nicaragua, and Panama. This is the case too with the seven chapters that Oviedo adds to *Book 11*.[46] The widespread paucity of information on flora and fauna divorced from Oviedo's experiences on Hispaniola and the Caribbean mainland confirms his difficulties in learning about New World nature beyond his direct reach. Oviedo would quickly have realized not only the impossibility but also the impracticality of using part 1 as a model for parts 2 and 3, as he had originally intended: such a comprehensive portrait of life would entail a great deal more knowledge of other regions in the Americas than what he has.

Although, ostensibly, the *Historia general* covers much more territory than the *Sumario* and the *Historia* (1535), then, its natural history continues to retain much the same profile. The relative continuity between the different phases of Oviedo's historiographical project means that he encountered similar editorial dilemmas throughout his writing career, although his solution to them changed over time. Recall from Chapter 3 that initially, in the *Sumario*, Oviedo struggles to strike a balance between showcasing the unique features of life on Hispaniola and the mainland, respectively, and acknowledging those characteristics they share. He divides the *Sumario* into two parts, corresponding to the island (chapters 2–8) and *tierra firme* (chapters 9–85); within this chapter division, flora and fauna function as subcategories pertaining to one or the other geography. Yet Oviedo does away with these clear-cut distinctions when they do not reflect the biological realities of life in the Caribbean: even though he includes the entry on yuca in the section devoted to Hispaniola, he describes a variety cultivated on *tierra firme* as well.[47] Oviedo's comments in the *Historia* (1535) regarding the future three-part work indicate that he originally planned to organize New World nature in the Monserrat manuscript according to similar criteria, with each phenomenon placed in its corresponding geography, as he had done in the *Sumario*.

Ultimately, however, Oviedo discards this narrative strategy for one in many ways reminiscent of his approach to natural history in the *Historia* (1535), a topic explored in Chapter 4. While Oviedo purports to explore Antillean nature at large in this work, Hispaniola is its true protagonist. At the same time, Oviedo dedicates chapters in the appropriate geographies to phenomena unique to those places; for example, he includes the *palo santo* in the books on San Juan (Puerto Rico) since, to his knowledge, it is found nowhere else in the Indies.

A similar logic dictates the organization of the Monserrat manuscript: just as Hispaniola signals nature's unity in the *Historia* (1535), so too does part 1 of the three-part work come to embrace the oneness of this other half of the world. Given that Oviedo's knowledge of New World nature continued to be restricted to the wider Caribbean during the last phase of his historiographical project, he would have observed the same overlapping patterns in the natural world as those that marked the *Historia* (1535). The Monserrat manuscript goes a step farther than its predecessor in its all-embracing treatment of nature in the Indies, however. The first part of the work brings together not only those flora and fauna that are common to Oviedo's Indies but also those which are different—buffalo and llamas taking their place alongside the pineapple and the iguana.[48] The Monserrat manuscript's natural history comprehends all New World nature's possibilities, not just those that are most ubiquitous.

Although the Monserrat manuscript of the *Historia* is inarguably Oviedo's magnum opus, it is not without precedent. Oviedo does not reinvent his natural history every time he writes it. Instead, he draws on his published works, the *Sumario* and the *Historia* (1535), as points of departure for his later explorations of this topic. How to acknowledge the inherent unity of New World nature while also paying homage to its diversity is a dilemma Oviedo faces repeatedly throughout his career; the Monserrat manuscript is yet another iteration of this challenge. In addition to his struggles to harness the complexities of a natural world that defies geographical and political borders, the limitations of Oviedo's eyewitness experiences, too, are fundamental in determining how he fashions New World nature. These are the concerns that continually plague Oviedo, which prompt him to adjust his initial plans for his post-1535 project soon after embarking on its composition. In the Monserrat manuscript, it is Nature with a capital N that reigns supreme and which Oviedo celebrates. The full force of its marvels can be felt in the complete, totalizing gaze that Oviedo offers to his readers, a veritable cornucopia of plants and animals.

NOTES

Preface

1. *Historia general de las Indias* (Monserrat), 321r (bk. 12, chap. 21). Unless otherwise noted, all translations are my own.

2. *Historia general* (Monserrat), 321r (bk. 12, chap. 21).

3. Consulting Muñoz's copy of the image, the Spanish scholar Jesús Carrillo Castillo observes that the drawing is much more elegant than others in *Book 12*, which he characterizes as being "clumsy and not very attractive." *Naturaleza e imperio: La representación del mundo natural en la "Historia general y natural de las Indias" de Gonzalo Fernández de Oviedo* (Madrid: Fundación Carolina, 2004), 321.

4. *Batallas y Quinquagenas*, ed. Juan Pérez de Tudela Bueso (Madrid: Real Academia de la Historia, [1535–1557] 1983), 3:257 [bat. 1, diálogo no clasificado] and 3:33 [bat. 1, quin. 4, diál. 7].

Introduction

1. Columbus believed that the first island on which he landed, which he named San Salvador, formed part of an archipelago off the mainland province of Magni, in today's southern China. Nicolás Wey Gómez, *The Tropics of Empire: Why Columbus Sailed South to the Indies* (Cambridge, MA: MIT Press, 2008), 11.

2. Gonzalo Fernández de Oviedo, *La historia general de las Indias con privilegio imperial. Fin dela primera parte dela general y natural historia delas indias yslas y tierra firme del mar oceano* (Seville: Juan Cromberger, 1535), bk. 2, chap. 7 [ix r]. The first edition of the *Historia* consists of part 1 (books 1–19) of what would later become a three-part work, as well as a book on shipwrecks (*Book 20*).

3. A *mozo de cámara* worked as a personal servant for a member of the palace. Real Academia Española, *Diccionario de la lengua española*, s.v. "mozo de cámara," https://dle.rae.es /mozo#EHl1WKA, accessed August 25, 2022.

4. *Historia* (1535), bk. 9, chap. 12 [xc r].

5. *Historia* (1535), bk. 10, chap. 4 [xciii v–xcv r].

6. *Historia* (1535), bk. 8, chap. 25 [lxxxvi v].

7. Brian W. Ogilvie, *The Science of Describing: Natural History in Renaissance Europe* (Chicago: University of Chicago Press, 2006); Alix Cooper, *Inventing the Indigenous: Local Knowledge and Natural History in Early Modern Europe* (Cambridge: Cambridge University Press, 2007); Paula Findlen, *Possessing Nature: Museums, Collecting, and Scientific Culture in Early Modern Italy* (Berkeley: University of California Press, 1994); Paula Findlen, ed., *Empires of Knowledge: Scientific Networks in the Early Modern World* (London: Routledge, 2018).

8. Antonio Barrera-Osorio, *Experiencing Nature: The Spanish American Empire and the Early Scientific Revolution* (Austin: University of Texas Press, 2006).

9. Arndt Brendecke, *The Empirical Empire: Spanish Colonial Rule and the Politics of Knowledge* (Berlin: De Gruyter, 2016), 12.

10. Daniela Bleichmar, *Visible Empire: Botanical Expeditions and Visual Culture in the Hispanic Enlightenment* (Chicago: University of Chicago Press, 2012).

11. Jorge Cañizares-Esguerra, *Nature, Empire, and Nation: Explorations of the History of Science in the Iberian World* (Redwood, CA: Stanford University Press, 2006); Andrés I. Prieto, *Missionary Scientists: Jesuit Science in Spanish South America, 1570–1810* (Nashville, TN: Vanderbilt University Press, 2011).

12. Several classic works on the role of Spain in the development of the sciences and in the dissemination of knowledge about New World nature include José María López Piñero, *Ciencia y técnica en la sociedad española de los siglos XVI y XVII* (Barcelona: Editorial Labor, S.A., 1979); José María López Piñero and María Luz López Terrada, *La influencia española en la introducción en Europa de las plantas americanas: 1493–1623* (Valencia: University of Valencia, 1997); and Raquel Álvarez Peláez, *La conquista de la naturaleza americana* (Madrid: Consejo Superior de Investigaciones Científicas, 1993). Given their vast scope and the great importance they accord to developments in these areas toward the end of the sixteenth century, they treat Oviedo's works only briefly.

13. Antonello Gerbi, *Nature in the New World: From Christopher Columbus to Gonzalo Fernández de Oviedo*, trans. Jeremy Moyle (Pittsburgh, PA: University of Pittsburgh Press, [1975] 2010), 131; Daniela Bleichmar, *Visual Voyages: Images of Latin American Nature from Columbus to Darwin* (New Haven, CT: Yale University Press, 2017), 20–21; María M. Portuondo, "Cosmography at the *Casa*, *Consejo*, and *Corte* During the Century of Discovery," in *Science in the Spanish and Portuguese Empires, 1500–1800*, ed. Daniela Bleichmar, Paula De Vos, Kristine Huffine, and Kevin Sheehan (Stanford, CA: Stanford University Press, 2009), 68.

14. The paragraphs that follow are not by any means an exhaustive survey of scholarship on Oviedo's life and writing. See Juan Bautista Avalle-Arce, "Oviedo a media luz," *Nueva Revista de Filología Hispánica* 29, no. 1 (1980): 138–151; Manuel Ballesteros Gaibrois, *Gonzalo Fernández de Oviedo* (Madrid: Fundación Universitaria Española, 1981); Juan Pérez de Tudela Bueso, "Vida y escritos de Gonzalo Fernández de Oviedo," in *Historia general y natural de las Indias*, by Gonzalo Fernández de Oviedo, 1:7–169 (Madrid: Ediciones Atlas, 1959); and José Amador de los Ríos, "Vida y escritos de Gonzalo Fernández de Oviedo y Valdés," in *Historia general y natural de las Indias*, by Gonzalo Fernández de Oviedo, 9–107 (Madrid: Real Academia de la Historia, 1851).

15. These include Enrique Álvarez López, "Plinio y Fernández de Oviedo," *Anales de ciencias naturales* (1940): 40–61; the introduction to a 1942 edition of the *Sumario*; and "La historia natural en Fernández de Oviedo," *Revista de Indias* 17, no. 69 (1957): 541–601.

16. The modern editions of the *Historia general*—edited by José Amador de los Ríos (1818–1878), Juan Natalicio González (1897–1966), and Juan Pérez de Tudela Bueso (1922–2004), respectively—likely contribute to this problem since they do not differentiate between the content of the 1535 edition and Oviedo's later additions and edits.

17. Antonello Gerbi perhaps most explicitly articulates the view of the *Historia general* as a composite text, writing, "We shall consider the *History* provisionally as a single unit, pointing out only, where opportune, certain shifts of accent or emphasis between the earliest and

the final books" (*Nature*, 216). Although the Italian scholar upholds this methodology as a solution to the challenges presented by the unfinished nature of the Monserrat manuscript—which Oviedo continued to emend and add to over time—nevertheless, the comment is applicable to his approach to the *Historia* (1535) as well. On the other hand, Alberto M. Salas confuses the two works: he describes the print edition of part 1 as embracing the nature worlds of New Spain and Río de la Plata, when in fact Oviedo did not include these geographies until later, in the Monserrat manuscript. Similarly, although Myers points to differences between the account of the pineapple in the *Historia* (1535) and the Monserrat manuscript, she cites uniquely from the latter (without indicating this to be the case). Other similar examples abound in studies on Oviedo's *Historia general*. Alberto Mario Salas, *Tres cronistas de Indias: Pedro Mártir de Anglería, Gonzalo Fernández de Oviedo, Fray Bartolomé de las Casas* (Mexico City: Fondo de Cultura Económica, 1986), 100; Kathleen Ann Myers, *Fernández de Oviedo's Chronicle of America: A New History for a New World* (Austin: University of Texas Press, 2007), 75–80.

18. Jesús Carrillo Castillo argues that these revisions represent a radical change in Oviedo's approach to New World nature that took place in 1541. See *Naturaleza e imperio*.

19. Oviedo writes little of his early life, though he mentions that he was born in Madrid and his parents were from Asturias.

20. *Historia general*, ed. Tudela Bueso, 1:30 [bk. 2, chap. 7].

21. "Asi alli se acabaron las esperanças que yo esperaua y me ofrescía esta vida. Por lo qual mi descontentamiento me lleuó fuera de España a peregrinar por el mundo." *Las memorias de Gonzalo Fernández de Oviedo*, ed. Juan Bautista Avalle-Arce (Chapel Hill: University of North Carolina Press, 1974), 2:543.

22. *Batallas*, ed. Tudela Bueso, 2:68 [bat. 1, quin. 2, diál. 20].

23. *Batallas y Quinquagenas*, ed. Juan Bautista Avalle-Arce (Salamanca: Ediciones de la Diputación de Salamanca, [1535–1557] 1989), 105–106.

24. *Batallas*, ed. Avalle-Arce, 453.

25. *Batallas*, ed. Tudela Bueso, 2:75 [bat. 1, quin. 2, diál. 20]; *Batallas*, ed. Avalle-Arce, 302.

26. "Tuvo quatro cosas en que ningún prínçipe de sus tiempos le hizo ventaja nj se sabe que le ygualasse en aquella sazón y eran éstas: cámara riqujsima, capilla, caça e caualleriza." *Catálogo real de Castilla*, ed. Evelia Ana Romano de Thuesen (PhD diss., University of California at Santa Barbara, June 1992), 1587–1588.

27. Carol Kidwell, *Pontano: Poet and Prime Minister* (London: Duckworth, 1991), 202.

28. Kidwell, *Pontano*, 283.

29. *Memorias*, 1:194 and 2:639–640; *Historia*, 1:14 [bk. 2, chap. 1].

30. Michael Mallett and Christine Shaw, *The Italian Wars, 1494–1559: War, State and Society in Early Modern Europe* (London: Pearson, 2012), 58.

31. *Historia general*: "Fué un muy excelente príncipe" (ed. Tudela Bueso, 5:220 [bk. 48, chap. 2]).

32. *Batallas*, ed. Tudela Bueso, 2:75 [bat. 1, quin. 2, diál. 20]. A *guardarropa* was the custodian of one's personal effects. Real Academia Española, *Diccionario de la lengua española*, s.v. "guardarropa," https://dle.rae.es/guardarropa, accessed August 25, 2022.

33. *Batallas*, ed. Avalle-Arce, 130.

34. Manuel Colmeiro, *Cortes de los antiguos reinos de León y de Castilla* (Madrid: Impresores de la Real Casa, 1884), 2:70–71; Oviedo, *Catálogo real*, 43.

35. *Batallas*, ed. Tudela Bueso, 1:313 [bat. 1, quin. 1, diál. 38].

36. *Colección de documentos inéditos relativos al descubrimiento, conquista y organización de los antiguas posesiones españolas de América y Oceanía* (Madrid: M. B. de Quirós, 1864–1884), 1:271 (hereafter cited as *CDI*).

37. A notary of mines and crime and judge of branding natives in Castilla del Oro.

38. *Historia general*, ed. Tudela Bueso, 3:207 [bk. 29, chap. 1]. *Veedor de minas y fundiciones* is an inspector of mines and smelting.

39. *Historia general*, ed. Tudela Bueso, 3:232 [bk. 29, chap. 8].

40. *Historia general*: "comenzarse la destruición de la tierra (a que ellos llamaban pacificación e conquistar)" (ed. Tudela Bueso, 3:234 [bk. 29, chap. 8]).

41. *Historia general*, ed. Tudela Bueso, 3:237 [bk. 29, chap. 9].

42. *Historia general*, ed. Tudela Bueso, 3:249 [bk. 29, chap. 11].

43. *Historia general*, ed. Tudela Bueso, 3:250 [bk. 29, chap. 11].

44. *Historia general*, ed. Tudela Bueso, 2:182 [bk. 18, prologue].

45. *Historia general*, ed. Tudela Bueso, 3:252 [bk. 29, chap. 11].

46. *Historia general*, ed. Tudela Bueso, 3:285 [bk. 29, chap. 20]

47. Gonzalo Fernández de Oviedo, *Dela natural hystoria delas Indias* (Chapel Hill: University of North Carolina Press, [1526] 1969), 4 [prologue].

48. Fernández de Oviedo, *Dela natural hystoria*, 2.

49. *Historia general*, ed. Tudela Bueso, 2:64 [bk. 13, chap. 9].

50. "Crónica general y natural historia destas Indias" (quoted in Tudela Bueso, "Vida," 116).

51. *Catálogo real*, 43.

52. *Catálogo real*, 45–46.

53. Tudela Bueso, "Vida," 128.

54. Tudela Bueso, "Vida," 119.

55. As a *procurador*, or advocate, Oviedo was appointed to speak on behalf of the island at court, defending its interests and complaining about any grievances. Real Academia Española, *Diccionario de la lengua española*, s.v. "procurador," https://dle.rae.es/procurador, accessed August 25, 2022. *Historia general*, ed. Tudela Bueso, 2:409 [bk. 24, chap. 7].

56. A military title indicating second-in-command. Real Academia Española, *Diccionario de la lengua española*, s.v. "comendador mayor," https://dle.rae.es/comendador?m=form#4WDTlEu, accessed August 25, 2022.

57. A post charged with the upkeep and administration of a fortress or castle. Real Academia Española, *Diccionario de la lengua española*, s.v. "alcaide," https://dle.rae.es/alcaide?m=form, accessed August 25, 2022.

58. *Cartas de la real audiencia de Santo Domingo (1530–1546)*, ed. Geraro Rodríguez Morel (Santo Domingo: Archivo general de la nación, 2007), 459.

59. *Memorias*, 1:343.

60. Daymond Turner, "The Aborted First Printing of the Second Part of Oviedo's *General and Natural History of the Indies*," *Huntington Library Quarterly* 46, no. 2 (1983): 109. Oviedo does not explain his involvement in the printing of the second edition, but he writes that he was in Salamanca in 1548, the same year to which he attributes its publication, erring by a year in his calculations. *Memorias*, 1:342.

61. *Memorias*: "de las casas de España ilustres e de nobles linajes e famosos caualleros" (2:599–600).

62. *Memorias*, 1:17–18.

63. *Memorias*, 1:19.

64. Turner, "Aborted," 118–119.

65. *Batallas*, ed. Avalle-Arce, 329.

Chapter 1

1. Paula Findlen, "The Renaissance of Science," in *The Oxford Illustrated History of the Renaissance*, ed. Gordon Campbell (Oxford: Oxford University Press, 2019), 378–481.

2. Juan Luis Vives, *Vives: On Education* (Lanham, MD: Rowman and Littlefield, [1531] 1971), 168.

3. *Vives*, 170–171.

4. Ann Blair, "Natural Philosophy," in *The Cambridge History of Science*, ed. Katharine Park and Lorraine Daston (Cambridge: Cambridge University Press, 2006), 3:392.

5. Martin Davies, "Making Sense of Pliny in the Quattrocento," *Renaissance Studies* 9, no. 2 (1995): 241.

6. Charles Nauert, "Humanists, Scientists, and Pliny: Changing Approaches to a Classical Author," *American Historical Review* 84, no. 1 (1979): 81.

7. Nauert, "Humanists," 77.

8. Paula Findlen, "Natural History," in *The Cambridge History of Science*, ed. Katharine Park and Lorraine Daston (Cambridge: Cambridge University Press, 2006), 3:440–441.

9. Edward Lee Greene, *Landmarks of Botanical History* (Redwood City, CA: Stanford University Press, 1983), 2:550.

10. Quoted in Findlen, "Natural History," 441.

11. Jerome J. Bylebyl, "The School of Padua: Humanistic Medicine in the Sixteenth Century," in *Health, Medicine and Morality in the Sixteenth Century*, ed. Charles Webster (Cambridge: Cambridge University Press, 1979), 341n21.

12. Gareth D. Williams, *Pietro Bembo on Etna: The Ascent of a Venetian Humanist* (Oxford: Oxford University Press, 2017), 142–143.

13. A letter from Bembo to Oviedo is dated April 20, 1538, and Oviedo wrote to Bembo on January 20, 1543. In addition to these two extant letters, Bembo mentioned Oviedo several times in his correspondence with Giovanni Battista Ramusio (1485–1557). Cro Stelio, "La correspondencia epistolar entre el Cardenal Bembo y Fernández de Oviedo: Implicaciones históricas," in *América y la España del Siglo XVI*, ed. Francisco de Solano and Fermin del Pino (Madrid: Instituto Gonzalo Fernández de Oviedo, 1982), 1:54.

14. Herrera contradicts Theophrastus's opinion that leaving the buds on the hazelnut tree will cause it to produce more fruit of a better quality. Herrera writes that it is only by removing the buds that a tree grows more fruit. Gabriel Alonso de Herrera, *Obra de agricultura*, ed. José Urbano Martínez Carrera (Madrid: Ediciones Atlas, [1513] 1970), 132 [bk. 3, chap. 10]. Examples of Herrera correcting earlier agronomic accounts abound in the work.

15. Nauert, "Humanists," 76; Carrillo Castillo, *Naturaleza*, 83. In the *Historia general*, Oviedo recalls information that he read for the first time in Pliny's *Historia naturalis*: "más de cuarenta años, que lo leí por primera vez y muchos después en Plinio" (ed. Tudela Bueso, 1:198 [bk. 6, chap. 39]). Carrillo Castillo identifies this passage as being redacted in 1541–1542, dating Oviedo's acquaintance with Pliny's work to the 1490s. Both Daymond Turner and Carrillo Castillo have identified the Florentine humanist Cristoforo Landino's translation of the *Historia naturalis* (1467) as the version of the text that Oviedo consulted. Daymond Turner, "Los libros de alcaide: La biblioteca de Gonzalo Fernández de Oviedo y Valdés," *Revista de Indias* 31 (1971), 177; Carrillo Castillo, *Naturaleza*, 86–87.

16. Davies, "Making Sense," 250.

17. In point of fact, Pliny dedicated his work to Titus. H. Rackham, introduction to *Natural History*, by Pliny the Elder, trans. H. Rackham (Cambridge, MA: Harvard University Press, 1938), viii.

18. *Historia general*, ed. Tudela Bueso, 1:14 [bk. 2, chap. 1].

19. Oviedo cites the work of the Dominican Antoninus of Florence's *Summa Historialis* (ca. 1459) to demonstrate the diverse opinions concerning the dedicatee of Pliny's *Historia naturalis*. According to Oviedo, the clergyman proposes yet another possible recipient of the work, Emperor Vespasian himself. *Historia general*, ed. Tudela Bueso, 1:14 [bk. 2, chap. 1].

20. The ways in which Oviedo used Pliny's work as a model for his own is a topic explored extensively in Enrique Álvarez López's "La historia natural en Fernández de Oviedo" (1957). Stephanie Merrim describes the relationship as follows: "In it [the *Natural History*] Oviedo found a means of ordering scientifically, seeing empirically, and perhaps even writing. . . . The *Natural History* set the Spaniard on his course and would lend authority and historical sanction to his findings." "The Apprehension of the New in Nature and Culture: Fernández de Oviedo's *Sumario*," in *1492–1992: Re/Discovering Colonial Writing*, ed. René Jara (Minneapolis, MN: Prisma Institute, 1989), 175.

21. Kathleen Ann Myers describes Oviedo's literary relationship with Pliny as one in which he "singles out Pliny's text as a primary model and sets out to modernize it and surpass it" ("Imitation, Authority, and Revision in Fernández de Oviedo's *Historia general y natural de las Indias*," *Romance Languages Annual* 3 [1991]: 526). As Myers points out, at times Oviedo seems to resent the absolute authority that Pliny enjoys. Yet Oviedo could not have intended to render "his own *Historia general y natural* more useful than that of his model" ("Imitation," 526). As Oviedo was well aware, he and his predecessor wrote about two very different parts of the world; Oviedo's *Historia general* could no sooner be more useful than Pliny's *Historia naturalis* than it could surpass it. Rather, Oviedo conceived of his natural history as a supplement to Pliny's work, incorporating information on New World flora and fauna that was not available to the Roman historian. Just as Oviedo encouraged others to expand on and correct his account as more became known of Spain's overseas territories, so too does he occasionally provide updates to entries in the *Historia naturalis* based on his experiences in the Caribbean.

22. *Dela natural hystoria*: "lo que oyó dijo a quién, y lo que leyó atribuye a los autores que antes que él lo notaron, y lo que él vido como testigo de vista acomuló en la sobredicha su *Historia*. Ymitando al mismo quiero yo" (3 [prologue]).

23. *Historia general*: "Las cuales no he sacado de dos mil millares de volúmines que haya leído como en el lugar suso alegado Plinio escribe . . . pero yo acumulé todo lo que aquí escribo, de dos mill millones de trabajos y nescesidades e peligros en veinte e dos años e más que ha que veo y experimento por mi persona estas cosas" (ed. Tudela Bueso, 1:11 [bk. 1, prologue]).

24. *Historia general*, ed. Tudela Bueso, 2:57 [bk. 13, chap. 2].

25. *Historia general*: "Y para este efeto busco testigos algunas veces en los auctores antiguos, para que me crean como autor moderno e que hablo de vista, contando estas cosas a los que están apartados destas nuestras indias, porque acá cuantos no fueren ciegos, las veen" (ed. Tudela Bueso, 2:63 [bk. 13, chap. 8]).

26. *Historia general*, ed. Tudela Bueso, 2:63 [bk. 13, chap. 8].

27. *Historia general*: "Y también podría ser que en estas partes se formasen e criasen de una manera, y en el Oriente e donde dicen Plinio y otros auctores que las hay, se engendrasen

de otra forma . . . porque natura en algunas partes hace en diferentes modos sus operaciones en un mismo género de criaturas" (ed. Tudela Bueso, 2:204 [bk. 19, chap. 8]).

28. In Oviedo's interactions with the indigenous populations of the New World he assumes yet a different attitude toward Pliny's *Historia naturalis*. Oviedo is struck to observe that the practice of making wine is widespread in certain communities since Pliny attributes these inventions to historical figures unknown to the natives. In such instances, Oviedo sets aside the writings of the ancients, arguing that the natives learned these skills directly from nature. Louise Bénat-Tachot, "La experiencia en el proceso cognitivo de las sociedades indígenas en la *Historia general y natural de las Indias* de Fernández de Oviedo," in *Pensamiento europeo y cultura colonial*, ed. Karl Kohut and Sonia V. Rose (Madrid: Iberoamericana, 1997), 253.

29. *Historia general*: "Le ayudemos [a Plinio] a escrebir lo que él no supo ni halló escripto en las partes australes e occidentales destas nuestras Indias ni en las otras regiones dellas" (ed. Tudela Bueso, 2:27 [bk. 12, prologue]).

30. *Batallas*: "El valeroso rey Ciro dezía que no era menos nesçesaria el arte de la guerra que la agricoltura" (ed. Avalle Arce, 440).

31. F. W. Kent, *Lorenzo de'Medici and the Art of Magnificence* (Baltimore: Johns Hopkins University Press, 2004), 117.

32. Mauro Ambrosoli, *The Wild and the Sown: Botany and Agriculture in Western Europe, 1350–1850* (Cambridge: Cambridge University Press, 1997), 21.

33. Marcus Tullius Cicero, *De natura deorum*, trans. Francis Brooks (London: Methuen, 1896), 2.60.151–152, https://oll.libertyfund.org/titles/539, accessed April 9, 2020.

34. *Historia general*, ed. Tudela Bueso, 1:226–228 [bk. 7, chap. 1]. Oviedo understands indigenous agronomic knowledge as necessary but incomplete, a topic that will be further explored in Chapter 4. Although colonists throughout the Americas depended on the natives for knowledge of the natural world, Oviedo maintains that only the former have the capacity to fully exploit their resources. He argues that any firsthand knowledge or skills the natives have acquired is insufficient; the written word provides access to time-tested, reliable agricultural practices.

35. Johanna Elizabeth Bauman, "Piero de' Crescenzi's *Liber ruralium commodorum*: Unearthing the Origins of the Pleasure Garden" (PhD diss., University of Virginia, 2000), 147.

36. Ambrosoli, *Wild*, 44.

37. Savastano Luigi, *Contributo allo studio critic degli Scrittori agrari italici: Pietro dei Crescenzi* (Acireale: Tip. Orario delle Ferrovie, 1922), 42–46.

38. Ambrosoli, *Wild*, 15; Bauman, "Piero," 3.

39. Luigi, *Contributo*, 47. No Spanish translations of the *Liber* existed, so Oviedo would have read the manual in Italian or Latin. Robert G. Calkins, "Piero de' Crescenzi and the Medieval Garden," in *Medieval Gardens*, ed. Elizabeth Blair MacDougall (Washington, DC: Dumbarton Oaks, 1986), 160–161.

40. Bauman, "Piero," 20–21.

41. Ambrosoli, *Wild*, 50.

42. Bauman, "Piero," 157.

43. Bauman, "Piero," 170.

44. *Batallas*: "una hermosa pieça e provechosa" (ed. Avalle-Arce, 113).

45. James S. Grubb, *Provincial Families of the Renaissance: Private and Public Life in the Veneto* (Baltimore: Johns Hopkins University Press, 1996), 154.

46. Quoted in Kent, *Lorenzo*, 112.

47. Kent, *Lorenzo*, 116.

48. Herrera, *Obra de agricultura*, 6.

49. *Historia general*, ed. Tudela Bueso, 1:8 [bk. 1, prologue].

50. *Historia general*, ed. Tudela Bueso, 4:299 [bk. 35, chap. 3].

51. Grubb, *Provincial Families*, 154.

52. *Naturalis historia*, trans. John Bostock (London: Taylor and Francis, [77–79 AD] 1855), bk. 19, chap. 19, http://data.perseus.org/citations/urn:cts:latinLit:phi0978.phi001.perseus-eng1:19.19. Although this edition takes the title *Natural History*, throughout the book I call Pliny's work by its more popular name, the *Historia naturalis*.

53. *Historia* (1535), lxxvi v [bk. 7, chap. 13]. It is possible that Oviedo confused the Schifanoia Palace (Esquiva Noya) with a different villa that Alfonso I (1505–1534) constructed on the island of the Belvedere, since the palace is not located on an island. Gerbi, *Nature*, 188.

54. Amanda Lillie, *Florentine Villas in the Fifteenth Century: An Architectural and Social History* (Cambridge: Cambridge University Press, 2005), 67–68.

55. Marie Luise Gothein, "Italy in the Time of the Renaissance and the Baroque Style," in *A History of Garden Art: From the Earliest Times to the Present Day*, ed. Walter P. Wright (Cambridge: Cambridge University Press, 2014), 1:219–220.

56. *Batallas*, ed. Tudela Bueso, 1:357; *Batallas*, ed. Avalle Arce, 351; *Batallas*, ed. Tudela Bueso, 2:133.

57. Fairchild D. Ruggles, *Gardens, Landscape, and Vision in the Palaces of Islamic Spain* (University Park: Pennsylvania State University Press, 2000), 17–18.

58. John H. Harvey, "Gardening Books and Plant Lists of Moorish Spain," *Garden History* 3, no. 2 (Spring 1975): 11.

59. Ruggles, *Gardens*, 20–21.

60. José Tito Rojo and Manuel Casares Porcel, "From the Andalusí Garden to the Andalusian Garden: Remnants and Re-Creation," in *Middle East Garden Traditions: Unity and Diversity*, ed. Michel Conan (Washington, DC: Dumbarton Oaks Research Library and Collection, 2007), 291.

61. Tito Rojo and Porcel, "From the Andalusí Garden," 290.

62. José Tito Rojo, "Permanencia y cambio en los jardines de la Granada morisca (1492–1571): Los jardines de los palacios nazaríes; La Alambra y el Generalife," in *Jardín y naturaleza en el reinado de Felipe II*, ed. Carmen Añón and José Luis Sancho (Madrid: Sociedad estatal para la conmemoración de los centenarios de Felipe II y Carlos V, 1998), 370.

63. Ambrosoli, *Wild*, 15.

64. "Cartas de Micer Andres Navagero a M. Juan Bautista Ramusio," in *Viajes de extranjeros por España y Portugal desde los tiempos más remotos hasta fines del siglo XVI*, ed. J. Garcia Mercadal (Aguilar: S.A. de ediciones, 1952), 879 and 886.

65. Mario Cermenati, "Un diplomatico naturalista del rinascimento, Andrea Navagero," *Nuovo Archivio Veneto* 25, no. 1 (1912): 176.

66. Amada Meneses, "Andrea Navagero, traductor de Gonzalo Fernández de Oviedo," *Revista de Indias* 18 (1958): 71. Navagero shared his translation with Ramusio, who later published it in *Libro secondo delle Indie Occidentali* (1534). The Venetian ambassador likely can be credited with beginning the correspondence between Oviedo and Ramusio. Alexandre Coello de la Rosa, "El proceso de escritura del *Sumario* (1526) de Gonzalo Fernández de Oviedo y Valdés en el seno de circulaciones y transferencias culturales con el humanismo italiano," *Pedralbes* 36 (2016): 149.

67. Mercadal, *Viajes*, 882.

68. Mercadal, *Viajes*: "Os maravillará que con las ocupaciones que tengo y con su importancia me ocupe en estas menudencias, que bien pueden llamarse así; pero no os asombre, porque aparte de mi cargo, en ninguna cosa pienso más ni con mayor gusto. Por tanto, ayudadme, querido Ramusio, con todo vuestro cuidado y diligencia en esto como en la cosa mayor y más importante que pudierais hacer por mí, y pensad que yo soy Epicuro y que he de pasar mi vida entera en los jardines" (883).

69. David Coffin, "The Gardens of Venice," in *Magnificent Buildings, Splendid Gardens*, ed. Vanessa Bezemer Sellers (Princeton, NJ: Princeton University Press, 2008), 198.

70. *Historia general*, ed. Tudela Bueso, 2:31 [bk. 12, chap. 5].

71. On this date, King Ferdinand of Spain and Louis XII of France signed the Treaty of Granada in which they divided the kingdom between them, alleging that King Frederick had betrayed the Christian faith by consorting with the Turks. Mallett and Shaw, *Italian Wars*, 58; Oviedo, *Catálogo*, 1587–1588.

72. Alessandro Tagliolini, *Storia del giardino italiano: Gli artisti, l'invenzione, le forme dall' antichità al XIX secolo* (Florence: La casa Husher, 1988), 73.

73. Valerio Giacomini, "L'orto botanico di Napoli," in *Orti botanici delle università italiane*, ed. Mario Bolli (Naples: University of Naples, 1965), 90.

74. *Historia general*, ed. Tudela Bueso, 1:239 [bk. 7, chap. 14].

75. Oviedo sustained multiple interactions with Pontano. See *Memorias*, 1:194 and 2:639–640; *Historia general*, ed. Tudela Bueso, 1:14 [bk. 2, chap. 1].

76. "Questi giardini poi avranno piante esotiche e rare, disposte con arte e con la debita cura." Giovanni Pontano, *I trattati delle virtù sociali: De Liberalitate, De Beneficentia, De Magnificentia, De Splendore, De Conviventia* (Italy: Edizioni dell'Ateneo, [1498] 1965), 277.

77. Pontano, *I trattati*: "L'uomo splendido, si procurerà dapprima dei vivai, e così pure stormi di pavoni, di fagiani, di pernici e di altri uccelli . . . e tutte le altre cose che siano capaci di dare all'animo un onesto piacere" (277).

78. Carl Caruso, *Adonis: The Myth of the Dying God in the Italian Renaissance* (London: Bloomsbury, 2013), 12.

79. Caruso, *Adonis*, 11.

80. Caruso, *Adonis*, 13.

81. *Historia general*, ed. Tudela Bueso, 1:243 [bk. 7, chap. 14].

82. *Historia general*, ed. Tudela Bueso, 1:248 [bk. 8, chap. 1].

83. Ambrosoli, *Wild*, 109.

84. Katherine A. McIver, *Cooking and Eating in Renaissance Italy: From Kitchen to Table* (Lanham, MD: Rowman and Littlefield, 2015), 78–79.

Chapter 2

1. *Historia general*: "de todas las que en mi vida he visto, e la más galana e polida" (ed. Tudela Bueso, 1:175 [bk. 6, chap. 15]).

2. *Historia general*: "Yo confieso que no habrá pintor que lo pinte, por lo que he dicho. . . . Para mí, yo la tengo por la más extremada en su plumaje o gentileza de todas las que yo he visto, y de la que más me he admirado. Ella es de tal artífice y mano hecha, que se puede y debe creer que no se le acabó el arte en ésta. Ni sus obras puede pintor ni escultor ni orador expresar tan al natural, ni perfectamente dar a entender ingenio humano, como ellas son" (ed. Tudela Bueso, 1:176 [bk. 6, chap. 15]).

3. Daymond Turner, "Forgotten Treasure from the Indies: The Illustrations and Drawings of Fernández de Oviedo," *Huntington Library Quarterly* 48, no. 1 (1985): 42.

4. I discuss the illustrations in the *Sumario*, the *Historia* (1535), and the Monserrat manuscript in Chapters 5, 6, and 7. Of the many who have commented on the visual dimension of Oviedo's writings on the Indies, the contributions of two more recent scholars stand out: Carrillo Castillo, *Naturaleza*; Myers, *Fernández*; Myers, "The Representation of New World Phenomena: Visual Epistemology and Gonzalo Fernández de Oviedo's Illustrations," in *Early Images of the Americas: Transfer and Invention*, ed. Jerry M. Williams and Robert E. Lewis (Tucson: University of Arizona Press, 1993), 183–213.

5. Cañizares-Esguerra, *Nature, Empire, and Nation*, 19.

6. Carrillo Castillo, *Naturaleza*, 277.

7. Chiyo Ishikawa, "*La llave de palo*: Isabel la Católica as Patron of Religious Literature and Painting," in *Isabel la Católica, Queen of Castile: Critical Essays*, ed. David A. Boruchoff (New York: Palgrave Macmillan, 2003), 103.

8. Peggy K. Liss, *Isabel the Queen: Life and Times* (Philadelphia: University of Pennsylvania Press, [1992] 2004), 249.

9. Mari-Tere Alvarez, "Artistic Enterprise and Spanish Patronage: The Art Market During the Reign of Isabel of Castile (1474–1504)," in *Art Markets in Europe, 1400–1800*, ed. Michael North and David Ormrod (Aldershot: Ashgate, 1998), 49.

10. César Pemán y Pemartín, *Juan van Eyck y España* (Cadiz: Museo Provincial de Bellas Artes de Cadiz, 1969), 30–31.

11. Jonathan Brown, *Painting in Spain, 1500–1700* (New Haven, CT: Yale University Press, 1998), 7.

12. J. Brown, *Painting*, 16.

13. Oviedo describes the responsibilities of the *mozo de cámara* as follows: "Estos moços de cámara por la mañana ivan adonde estava el que tenía la ropa e llaves de la cámara, e limpiavan las calças, e borzeguís, e alcorques e vestidos que el prínçipe truxo el día antes, e cogido e limpio todo muy bien, lo ponían sobre una mesa, de donde aquellas ropas o otras que allí se les davan, las llevavan tras el camarero e ivan a vestir al prínçipe como ya está dicho. E en todos los tiempos del día e noche, hasta que Su Alteza se quería acostar, están donde su persona real está . . . en espeçial dos dellos a quien cabía la guarda de aquel día, para que con aquéllos Su Alteza embía a mandar lo que es servido o van con algún recabdo algunas vezes fuera de palaçio a dezir o hazer lo que el prínçipe es servido." *Libro de la cámara real del príncipe don Juan*, ed. Santiago Fabregat Barrios (Valencia: University of Valencia, [1548] 2006), 103–104. J. Brown, *Painting*, 19.

14. Sereno and the Alcaide both recall having arrived at the court of the Catholic Monarchs in 1490. *Batallas*, ed. Tudela Bueso, 2:151 [bat. 1, quin. 2, diál. 44] and 3:140 [bat. 1, quin. 4, diál. 32].

15. J. Brown, *Painting*, 7.

16. As Jonathan Brown explains, lack of interest in the technical execution of Flemish paintings was likely due to the fact that much of the art to reach Spain was imitation, and therefore had "already lost the veneer of refinement and sophistication" of the original masterpieces (*Painting*, 7).

17. C. D. Dickerson, "Beginnings in Castile," in *Alonso Berruguete: First Sculptor of Renaissance Spain*, ed. C. D. Dickerson III and Mark McDonald (Washington, DC: National Gallery of Art, 2019), 13.

18. Jessica Sue Weiss, "Isabel of Castile, Flemish Aesthetics, and Identity Construction in the Works of Juan de Flandes" (PhD diss., University of Texas at Austin, 2014), 65; Ishikawa, "*La llave*," 104.

19. Joaquin Yarza Luaces, *Los reyes católicos: Paisaje artístico de una monarquía* (Madrid: Editorial Nerea, 1993), 90.

20. Jessica Sue Weiss, "Juan de Flandes and His Financial Success in Castile," *Journal of Historians of Netherlandish Art* 11, no. 1 (Winter 2019): 2; Weiss, "Isabel," 12–13. Evidence suggests that Antonio Inglés was at the Castilian court between 1489 and 1490, where he arrived as part of the English delegation for the marriage negotiations of Catharine of Aragon to the Prince of Wales. He created portraits of the Castilian royal children but continued to work primarily in the employ of the English Crown. Weiss, "Juan de Flandes," 26n7.

21. Liss, *Isabel*, 296.

22. Ximenez de Salas, "Jaime Serrat, pintor del Príncipe Don Juan," *Archivo español de arte y arqueología* 12, no. 36 (1936): 269.

23. *Libro*: "que ambos son gentiles ofiçios entre los mecánicos e artesanos; e eran del príncipe mi señor, la mayor parte del tiempo ocupados, porque era muy inclinado a se bordar e vestir ricamente, e eran favoresçidos e bien pagados e aposentados" (164).

24. Ximenez de Salas, "Jaime Serrat," 269.

25. Ishikawa, "*La llave*," 104.

26. Yarza Luaces, *Los reyes*, 88–90.

27. Ishikawa, "*La llave*," 104.

28. Yarza Luaces, *Los reyes*, 360.

29. Erwin Panofsky, *Renaissance and Renascences in Western Art* (New York: Harper and Row, 1972), 26–27.

30. J. Brown, *Painting*, 29.

31. J. Brown, *Painting*, 28–29.

32. J. N. H. Lawrance, "The Spread of Lay Literacy in Late Medieval Castile," *Bulletin of Hispanic Studies* 62, no. 1 (1985): 87.

33. Christoforo Landino, *Comento di Christophoro Landino Fiorentino sopra la comedia di Danthe Alighieri poeta Fiorentino* (Bressa: Boninum de Boninis di Raguxi, 1487), 12 [prologue], https://archive.org/details/McGillLibrary-rbsc_incunabula_Divina-Commedia-di-Dante-Alighieri_1487-20449/page/n3/mode/2up, accessed October 21, 2022.

34. The only exception to this dearth of Greek art in the early modern period was the unearthing of the *Lacoön and His Sons* in 1506. The sculpture appears to be the same that Pliny describes in the *Historia naturalis*, which he refers to as superior to any other Greek statue or painting. The discovery of the *Lacoön* cemented the Roman historian's authority in the Renaissance as an art historian. Sarah Blake McHam, *Pliny and the Artistic Culture of the Italian Renaissance* (New Haven, CT: Yale University Press, 2013), 10–12.

35. McHam, *Pliny*, 3.

36. McHam, *Pliny*, 80.

37. *Naturalis historia*, bk. 35, chap 5.

38. *Batallas*: "Habés de saber que don Alonso solía decir algunas veces a los que le servían, que no valía nada el page a los otros hombres sin alguna especial habilidad; e que una de las principales era ser jentiles dibujadores. Siguióse que, estando una noche el caballero escribiendo de su mano, daba la sombra, que la candela dél ponía, en una pared blanca desviada de un lado donde el page estaba aguardando a su señor para despabilar las belas e hacer

lo que le mandase. E como era invierno, había una chimenea con fuego. . . . E como tenía gana de saber dibujar, tomó un carbón, e acullá en la pared donde la cara y luengas nariçes del caballero en la sombra de la pared se mostraban, se començó por el perfil de la sombra a poner la boca o misma forma de la sombra. E en esto embebecido, e casi fecha una cabeza grande e monstruosa pintura de su dibujo, el caballero bolvió acaso la cabeza e vió lo quel page hacía. E el page le dijo: 'Estése quedo vuesamerced, que ya acabo.' Pues, el caballero, riyéndose dijo: '¿Por qué no me lo decís antes, necio? Pues que viendo yo ese dibujo, no me pueden parescer grandes mis narices midiéndolas con ese espejo: acábala, acábala, que ese fue el principio de la pintura en el mundo.' Y fue berdad, que el caballero envió después a Flandes aquel page e gastó con él dineros para que aprendiese a pintar, e salió gentil pintor" (ed. Tudela Bueso, 3:64–65 [bat. 1, quin. 4, diál. 17]).

39. *Naturalis historia*, bk. 35, chap. 43.

40. *Naturalis historia*, bk. 35, chap. 44.

41. *Naturalis historia*, bk. 35, chap. 36.

42. *Naturalis historia*, bk. 35, chap. 36.

43. *Batallas*: [Sereno:] "Pienso yo, pues, que aquello es mejor pintura que más semejante es a las cosas naturales" (ed. Avalle-Arce, 94 [bat. 2, quin. 1, diál. 4]).

44. *Batallas*, ed. Avalle-Arce, 105 [bat. 2, quin. 1, diál. 5].

45. *Batallas*: [Alcaide:] "E avn lo que he dicho no es de menos exçelencia e arte que aquella escultura del olympico Júpiter que hizo Phidia, ni las esculturas de Pratixeles, ni de quanto labraron Scopa, Briante e Lisímaco, e trabajaron en aquella memorable sepoltura del rrey Mauseolo, ni de lo que pintaron Apeles e Protógenes. . . . E nuestro Berruguete en todas las que naturalmente se pueden ymitar, e como conuiene a la encarnaçión o espíritu e semejança de su perssona que quiere contrahazer" (ed. Avalle-Arce, 105 [bat. 2, quin. 1, diál. 5]).

46. *Batallas*: "Así me marauillo yo de vos, seyendo sabio, que pensés que he yo de tener en más la pintura que no veo della sino las palabras de los que la loan e ensalçan, que la que mirándola me paresçe tan proprio lo que rrepresentta, viendo por la espiriençia que ordinariamente todos los artes cresçen e se avmentan e se perfeçionan con el vso y exerçiçio" (ed. Avalle-Arce, 105 [bat. 2, quin. 1, diál. 5]).

47. Carrillo Castillo, *Naturaleza*, 195.

48. *Naturalis historia*, bk. 35, chap. 36.

49. *Historia general*, ed. Tudela Bueso, 2:7 [bk. 10, prologue].

50. *Batallas*: "Quanto más tarda de se acabar esta materia (destos diálogos), tanto más me paresçe que abunda la ocasión de nos detener en ellos, e tanto más difíçil me muestra el tiempo la conclusión, si no se vsase el rremedio de que vsauan algunos pintores antiguos. Y era que quando no les bastaua el arte se aprouechauan de vna cautela discreta: en pintar las ymágines de espaldas, queriendo dar a entender que aquellas eran más hermosas que las que mostrauan magnifiestas. Y tanbién es a nuestro propósito lo que dize Plinio de aquel gran pintor Timantho, que pintando a muchos en tristeza, cubrió, al que más triste avía de estar, la cara, que era padre de Ephigenia, que avía de ser sacrificada. E por tanto, dize en loor deste pintor que en todas sus obras se vehía más de lo que avía pintado. Así, pues, avés de entender que en lo que yo dixere de la infelicidad del serenísimo y exçelente señor don Fernando de Aragón, Duque de Calabria, es mucho mayor su pérdida e trabajo que yo os lo sabré pintar nj dezir, sin comparaçión" (ed. Avalle-Arce, 130 [bat. 2, quin. 4, diál. 2]).

51. *Batallas*: "Los versos del Duque son más profundos de lo que paresçe que suenan, ymitando a las obras de aquel pintor Timantho" (ed. Avalle-Arce, 141 [bat. 2, quin. 4, diál. 2]).

52. Michael Baxandall, *Giotto and the Orators: Humanist Observers of Painting in Italy and the Discovery of Pictorial Composition, 1350–1450* (Oxford: Clarendon Press, 1971), 97.

53. Baxandall, *Giotto*, 58.

54. Baxandall, *Giotto*, 53.

55. Oviedo quotes the line, "ningún bueno muere mal, e ningún malo bien" from *De remediis* in the *Historia general* (ed. Tudela Bueso, 4:362 [bk. 41, chap. 3]). Unlike Petrarch's sonnets, which Oviedo includes in their original Italian, he cites from *De remediis* in Spanish; likely, he had access to one of the seven Spanish editions of the work published in Spain before 1535. These editions include one published by Diego de Gumiel in Valladolid in 1510 and others by Jacob Cromberger (Seville, 1513), Juan Varela (Seville, 1516), Georges Coçi (Zaragoza, 1523), and Juan Varela (Seville, 1524 and 1534). Turner, "Los libros," 175.

56. Petrarch, *Petrarch's Remedies for Fortune Fair and Foul: Book 1*, trans. Conrad H. Rawski (Bloomington: Indiana University Press, [1354–1366] 1991), 126–128.

57. Meredith J. Gill, *Augustine in the Italian Renaissance: Art and Philosophy from Petrarch to Michelangelo* (Cambridge: Cambridge University Press, 2005), 114.

58. *Historia general*: "tan loable ocupación como es ver con los ojos corporales lo que hay en esta composición a ellos visible" (ed. Tudela Bueso, 1:8 [bk. 1, prologue]).

59. *Batallas*: "Fue don Vgo muy enamorado de vna muy hermosa e generosa dama, que yo vi muchas vezes en la çibdad de Nápoles. E avía procurado con diuerssos y exçelentes pintores que la contrahiziesen e pintassen al natural. E ninguno la hazía tan perfecta como él la quisiera, e con todo esso guardaua todos aquellos rretractos. . . . E desta manera thenía en su cámara, con mucha veneraçión todas aquellas figuras, por su deuoçión e passatiempo, vnas alegres, otras sañosas, vnas tocadas e bien ataviadas, otras en cabello e en diferentes meneos. E rriyéndome yo vn día con él sobrestos rretractos e pinturas, e mirándolos, e diziendo: 'Esta figura le paresçe más que aquésta, y aquélla que esotra, y estotra que la otra,' le pregunté que qué acordaua de hazer de aquellas gentes o mugeres insensibles. E don Vgo me rrespondió con vn gran sospirar y dixo: 'Sabe Dios que quanto querría yo más la biua que las pintadas, pero gran delito sería tocar en éstas, ni que se dexassen de estimar, porque se hizieron por ymitar la que está adentro en mj ánima.' E diziendo esto poníase la mano en el pecho" (ed. Avalle-Arce, 420 [bat. 2, quin. 2, diál. 37]).

60. *Batallas*: "Es mejor lo natural que lo artifiçial, y la diferençia es como del muerto al biuo, y quanto de la verdad a la mentira, y quanto de la sombra a lo çierto" (ed. Avalle-Arce, 420 [bat. 2, quin. 2, diál. 37]).

61. Oviedo, *Catálogo real*, 1581.

62. Oviedo, *Catálogo real*, 1591.

63. *Batallas*, ed. Tudela Bueso, 2:68 [bat. 1, quin. 2, diál. 20].

64. *Batallas*: "pintor exçelente" (ed. Avalle-Arce, 105 [bat. 2, quin. 1, diál. 5]).

65. Martin Kemp, *Leonardo da Vinci: The Marvellous Works of Nature and Man* (Oxford: Oxford University Press, [1981] 2006), 22.

66. Kemp, *Leonardo*, 25.

67. Kenneth Clark, introduction to *Leonardo da Vinci: Nature Studies from the Royal Library at Windsor Castle*, by Carlo Pedretti (New York: Johnson Reprint, 1981), 9.

68. *Batallas*, ed. Avalle-Arce, 105–106 [bat. 2, quin. 1, diál. 5].

69. Kemp, *Leonardo*, 76; *Batallas*, ed. Avalle-Arce, 105–106 [bat. 2, quin. 1, diál. 5]; Carrillo Castillo, *Naturaleza*, 204.

70. Leonardo da Vinci, *Treatise on Painting*, trans. A. Philip McMahon (Princeton, NJ: Princeton University Press, 1956), 9.

71. *Historia general*: "Los ojos son mucha parte de la información destas cosas" (ed. Tudela Bueso, 1:268 [bk. 8, chap. 30]).

72. *Batallas*, ed. Avalle-Arce, 311 [bat. 2, quin. 2, diál. 38].

73. *Batallas*: "Era tan exçelente la obra que todo lo que tengo dicho lo endendiera vn mediano entendimiento" (ed. Avalle-Arce, 311 [bat. 2, quin. 2, diál. 38]).

74. *Batallas*: "Fue vno de los famosos e sotiles ombres de su ofiçio en aquella sazón de toda Italia" (ed. Avalle-Arce, 311 [bat. 2, quin. 2, diál. 38]).

75. Clifford M. Brown and Sally Hickson, "Caradosso Foppa (ca. 1452–1526/27)," *Arte Lombarda* 119, no. 1 (1997): 12.

76. Barbara Furlotti and Guido Rebecchini, *The Art of Mantua: Power and Patronage in the Renaissance*, trans. A. Lawrence Jenkens (Los Angeles: J. Paul Getty Museum, 2008), 95.

77. Francis Ames-Lewis, *Isabella and Leonardo: The Artistic Relationship Between Isabella d'Este and Leonardo da Vinci, 1500–1506* (New Haven, CT: Yale University Press, 2012), 32, 75–82.

78. David Chambers, Jane Martineau, and Rodolfo Signorini, "Mantegna and the Men of Letters," in *Andrea Mantegna*, ed. Jane Martineau (New York: Metropolitan Museum of Art, 1992), 11.

79. Keith Christiansen, *The Genius of Andrea Mantegna* (New York: Metropolitan Museum of Art, 2009), 7; Chambers, Martineau, and Signorini, "Mantegna," 19–20.

80. Christiansen, *Genius*, 40. Francesco Gonzaga described the paintings as "almost alive and breathing so that the subject seems not represented but actually to exist" (Christiansen, *Genius*, 39).

81. Isabella d'Este commissioned Mantegna to paint three canvases for her *studiolo*. The artist completed the first, *Parnassus*, by July 1497, several years before Oviedo visited Mantua. Mantegna would not finish the second, *Minerva Chasing the Vices from the Garden of Virtue*, until 1502, and the third remained uncompleted upon his death in 1506. Furlotti and Rebecchini, *Art of Mantua*, 96–99. *Batallas*: "Andrea Manteña, que era otro Leonardo de Avince" (3:261, ed. Tudela Bueso [bat. 1, diálogo no clasificado]).

82. *Batallas*, ed. Avalle-Arce, 81 [bat. 2, quin. 2, diál. 41].

83. *Batallas*: "E todo lo ques dicho en tanta perfiçión que era cosa marauillosa el grandísimo primor e sotileza de todas aquellas figuras e actos que aquel mundo rrepresentaua, tanto que era menester que con buena vista e atençión y muy despaçio se considerasse cada cosa, y avn tener buen conosçimiento del arte de la pintura e sentirla para gozar de tanta moltitud de primores" (ed. Avalle-Arce, 81 [bat. 2, quin. 2, diál. 41]).

84. *Batallas*, ed. Tudela Bueso, 3:260–262 [bat. 1, diálogo no clasificado].

85. *Batallas*: "Y el Leonardo dixo: 'esta es la cosa del mundo que hasta hoy he visto que más me haya maravillado. Y si no se lo viera cortar, yo no creyera que hombre podía hacer cosa tan sutil con solas las tijeras y sin dibujo alguno'" (ed. Tudela Bueso, 3:260 [bat. 1, diálogo no clasificado]).

86. *Batallas*: "tablas de Martinus e de otros grandes barones del buril" (ed. Tudela Bueso, 3:261 [bat. 1, diálogo no clasificado]).

87. Alan Shestack, introduction to *Complete Engravings of Martin Schongauer* (New York: Dover, 1969), v.

88. Pamela H. Smith, *The Body of the Artisan: Art and Experience in the Scientific Revolution* (Chicago: University of Chicago Press, 2004), 60.

89. David Landau and Peter Parshall, *The Renaissance Print, 1470–1550* (New Haven, CT: Yale University Press, 1994), 50.

90. *Batallas*, ed. Tudela Bueso, 3:261 [bat. 1, diálogo no clasificado].

91. Dickerson, "Beginnings," 13.

92. Dickerson, "Return to Spain, *Pintor del Rey*, and Learning to Sculpt," in *Alonso Berruguete: First Sculptor of Renaissance Spain*, ed. C. D. Dickerson III and Mark McDonald (Washington, DC: National Gallery of Art, 2019), 38.

93. Dickerson, "Return," 38.

94. *Batallas*, ed. Avalle-Arce, 105 [bat. 2, quin. 1, diál. 5].

95. Richard L. Kagan, "The Quest for Status and Wealth," in *Alonso Berruguete: First Sculptor of Renaissance Spain*, ed. C. D. Dickerson III and Mark McDonald (Washington, DC: National Gallery of Art, 2019), 62; Manuel Arias Martínez, "Master of the Retablo," in *Alonso Berruguete: First Sculptor of Renaissance Spain*, ed. C. D. Dickerson III and Mark McDonald (Washington, DC: National Gallery of Art, 2019), 86.

96. *Batallas*, ed. Avalle-Arce, 105 [bat. 2, quin. 1, diál. 5]. Although Berruguete was a salaried court painter until 1532, as early as 1523 he received permission to absent himself from court. Notwithstanding the prestige of his title, Berruguete was often tasked with menial jobs that doubtlessly did not meet his expectations. He sought to establish an independent workshop with more freedom to accept outside commissions. Kagan, "Quest," 57.

97. Pliny the Elder, *Naturalis historia*, bk. 10, chap. 3; Isidore of Seville, *The Etymologies of Isidore of Seville*, ed. Stephen A. Barney (Cambridge: Cambridge University Press, 2006), 264 [bk. 12, chap. 7].

98. *Batallas*, ed. Avalle-Arce, 105 [bat. 2, quin. 1, diál. 5].

99. *Batallas*: "Ésta era vna de las más perfectas pinturas que pintada fue de aquel famoso Berruguete en toda su vida" (ed. Avalle-Arce, 105 [bat. 2, quin. 1, diál. 5]).

100. Pliny, *Naturalis historia*, bk. 34, chap. 9.

101. Pliny, *Naturalis historia*, bk. 34, chap. 9.

102. Petrarch, *Petrarch's Remedies*, 133.

103. Stephen K. Scher, "An Introduction to the Renaissance Portrait Medal," *Perspectives on the Renaissance Medal*, ed. Stephen K. Scher (New York: Garland, 2000), 3.

104. Scher, "Introduction," 3.

105. John Cunnally, *Images of the Illustrious: The Numismatic Presence in the Renaissance* (Princeton, NJ: Princeton University Press, 1999), 35.

106. Liana de Girolami Cheney, "The Impresa in the Italian Renaissance," in *Companion to Emblem Studies*, ed. Peter M. Daly (New York: AMS Press, 2008), 252.

107. Carrillo Castillo, *Naturaleza*, 208.

108. *Batallas*, ed. Tudela Bueso, 3:257 [bat. 1, diálogo no clasificado] and 3:33 [bat. 1, quin. 4, diál. 7].

109. *Batallas*, ed. Avalle-Arce, 311–312 [bat. 2, quin. 2, diál. 38].

110. *Batallas*: "Las lágrimas represadas / más duelen que las mostradas" (ed. Avalle-Arce, 312 [bat. 2, quin. 2, diál. 38]).

111. *Batallas*: “Dize que morirá sin esperar que le puedan llorar otros, nj él pueda tanpoco llorarse a sí nj su pena” (ed. Avalle-Arce, 311 [bat. 2, quin. 2, diál. 38]).

112. *Batallas*: “ninguno de seruitud jamás se dolió, nj de muerte, quanto el de la libertad e de al vida que otrié no le quitó” (ed. Avalle-Arce, 312 [bat. 2, quin. 2, diál. 38]). Petrarch, “The Triumph of Death,” in *The Sonnets, Triumphs, and Other Poems of Petrarch* (London: George Bell and Sons, 1879), 376.

113. *Batallas*, ed. Avalle-Arce, 312 [bat. 2, quin. 2, diál. 38].

Chapter 3

1. Angela Pérez Ovejero, introduction to *Libro secondo delle Indie Occidentali*, by Gonzalo Fernández de Oviedo (Rome: Bulzoni, 1990), 18 and 33; Gerbi, *Nature*, 166. Ramusio included the Italian translation of the *Sumario*, along with Pietro Martire d’Anghiera’s *De orbo novo decades* (*Decades of the New World*) and Francisco de Xerez’s account of the conquest of Peru, in the third volume of *Della navigationi et viaggi* (1556). Pietro Bembo also included a summary of the *Sumario* in the sixth book of his *History of Venice* (written between 1534 and 1538, published in 1551).

2. Oviedo makes various allusions in the *Historia general* to having documented life at the court of the Catholic Monarchs, where he served as a *mozo de cámara* for Prince John (1490–1497). It is possible that he acquired this habit from his father’s diarist inclinations; in *Batallas*, Oviedo explains that his father, “desiring to know and understand what he saw . . . recorded many memories of these occurrences” (ed. Tudela Bueso, 3:15 [bat. 1, quin. 4, diál. 2]). In the *Historia* (1535), Oviedo mentions having recorded in his diary (“memoriales”) several events from the last decade of the fifteenth century, among them the conquest of Granada and the parade of six Antillean natives that Christopher Columbus presented to the king and queen in Barcelona upon the Genovese’s return from the Caribbean (1535, ix r [bk. 2, chap. 7]).

3. In the *Catálogo real*, Oviedo reminisces on the origins of this project: “Lo qual yo començé a copilar el año de mill e quinientos e cinco de la natiuidad del Señor, conosçiendo quel Serenísimo e Cathólico Rey don Fernando, quinto de tal nombre, vuestro abuelo, desseaua una copilación semejante en que se tractasse lo que en esta primera parte e en otras dos espero continuar” (43). Evelia Ana Romano de Thuesen describes the work as follows: “[It] concerns the genealogy of the reigning houses of Spain and Europe and their relationships, focusing on the ascendants of Charles V. At the same time it develops the chronology of empire since the very first emperor, during the Roman era, Cayo Julio Ceasar and concludes with the account of Popes, since Saint Peter.” Evelia Ana Romano de Thuesen, abstract to *Catálogo real de Castilla*, by Gonzalo Fernández de Oviedo, ed. Evelia Ana Romano de Thuesen (PhD diss., University of California at Santa Barbara, June 1992), viii.

4. Richard L. Kagan, *Clio and the Crown: The Politics of History in Medieval and Early Modern Spain* (Baltimore: Johns Hopkins University Press, 2009), 27 and 63–64. Charles V’s attitude toward history embodies a trend of increasingly narrowed historiographical interests on the part of Castile’s monarchs, beginning around 1284 and culminating with the accession of the Catholic Monarchs in 1474. The king, like his grandparents, moved away from the broader historical perspectives articulated in general chronicles, such as Alfonso X’s *Estoria de Espanna*, embracing instead individualized, king-centered royal chronicles. For this reason, when the Cortes of Castile began urging Charles V to appoint a chronicler to update Alfonso X’s *Estoria de Espanna* for publication, in 1523, initially the king was reluctant to do so. It was not until 1539 that he finally acceded to these requests, appointing Florián Ocampo

as royal chronicler with specific instructions to prepare a print manuscript of the *Estoria*. Although this new edition would be published several years later, in 1543, it was at the instigation of a private bookseller, not Charles V. Instead, the monarch's historical preferences tended toward "his own history together with the ongoing discovery and conquest of the Indies" (*Clio*, 64).

5. *Dela natural hystoria*, 4 [prologue].

6. *Dela natural hystoria*, 3–4 [prologue].

7. *Dela natural hystoria*, 103.

8. *Dela natural hystoria*: "escribir en limpio" (103).

9. Myers, "Imitation," 523n1. Antonello Gerbi agrees with this assessment. He contests the notion that the *Sumario* is simply a "précis of the *History*," given that so little of the latter had yet been written (*Nature*, 214). Oviedo adds still other sections from the *Sumario* to the post-1535 Monserrat manuscript.

10. Gerbi, *Nature*, 215. Gerbi arrives at this conclusion partially based on his observation that "many passages from the *Summary* were simply copied out in the corresponding places in the *History*. Some were developed and altered; a few others were omitted altogether" (214–215). I would argue that these editorial decisions, far from indicating that the *Sumario* and the *Historia general* should be considered as one and the same, point to the development of Oviedo's natural history project over time. Barrera-Osorio, *Experiencing*, 105.

11. Andrés I. Prieto, "Classification, Memory, and Subjectivity in Gonzalo Fernández de Oviedo's *Sumario de la natural historia* (1526)," *MLN* 124, no. 2 (2009): 329.

12. Merrim, "Apprehension"; Jeremy Paden, "The Iguana and the Barrel of Mud: Memory, Natural History, and Hermeneutics in Oviedo's *Sumario de la natural historia de las Indias*," *Colonial Latin American Review* 16, no. 2 (2007): 217.

13. Antonio Sánchez-Jiménez addresses this topic in "Memoria y utilidad en el *Sumario de la natural historia de las Indias* de Gonzalo Fernández de Oviedo" (*Colonial Latin America Review* 13, no. 2 [2004]: 263–273), in which he explores the possible mnemonic techniques Oviedo used to recall and organize information on the natural world. Jeremy Paden ("Iguana") and Andrés Prieto ("Classification") argue that memory is an organizing principle of the *Sumario*, which furthermore endows the text with an ideological bent and puts it in contact with classical tradition.

14. Edmundo O'Gorman attributes Oviedo's decision to write the *Sumario* to a sudden realization of the possibilities inherent in the subject of the Americas and a desire to claim its novelties for his own; as he writes, "En un moment feliz se percataría Oviedo de que América andaba en busca de autor." O'Gorman, *Cuatro historiadores de Indias, siglo XVI: Pedro Martír de Anglería, Gonzalo Fernández de Oviedo y Valdés, Bartolomé de las Casas, Joseph de Acosta* (Mexico City: Secretaría de Educación Pública, 1972), 46. While making a name for himself as a chronicler of the Indies was likely one of Oviedo's objectives in composing the work, I argue that his motivations were diverse and, in part, tied to the historical circumstances in which he found himself.

15. As Jesús Carrillo Castillo states, "No obvious precedent for the description of natural phenomena existed in Spanish literary and academic contexts." "'The World Is Only One and Not Many': Representation of the Natural World in Imperial Spain," in *Spain in the Age of Exploration, 1492–1819*, ed. Chiyo Ishikawa (Lincoln: University of Nebraska Press, 2004), 144.

16. *Dela natural hystoria*: "el qual major que otro autor en lo que toca a la natural ystoria" (3 [prologue]).

17. *Dela natural hystoria*: "dixo lo que pudo ver y entendió de semejantes materias. Esta fue la opinión de Plinio" (3 [prologue]).

18. Álvaro Baraibar, "Estudio preliminar," in *Sumario de la natural historia de las Indias*, by Gonzalo Fernández de Oviedo (Madrid: Iberoamericana Vervuert, [1526] 2010), 17–18. Originally *tierra firme* referred to the northern coast of South America. As the Spanish continued to encounter new coastline along the Atlantic Ocean, however, the Crown sought to create different territorial jurisdictions. Its concessions to Pedrarias Dávila, on July 27, 1513, mark the first time that the Caribbean coasts of southern Central America and northwestern South America were called Castilla del Oro. The title that gave Pedrarias jurisdiction over Castilla del Oro included concessions formerly awarded to Alonso de Hojeda and Diego de Nicuesa. The latter had been given Veragua, which consisted of all known land west of Urabá on the mainland as far as Honduras, while the former was given a territory stretching east from the province of Urabá to the Cabo Isleos (Cabo Codera, Venezuela). Ongoing litigation with Columbus's heirs excluded Veragua from Pedrarias's authority, though the title he received did not specify its borders except to say that Darién was no longer part of this territory. Provinces that Oviedo references in the *Sumario*—such as Cueva, Darién, Urabá, and Coiba—form part of Pedrarias's governance. Carl Ortwin Sauer, *The Early Spanish Main* (Berkeley: University of California Press, 1969), 1, 169, and 247; Rolena Adorno and Patrick Charles Pautz, *Álvar Núñez Cabeza de Vaca: His Account, His Life, and the Expedition of Pánfilo de Narváez* (Lincoln: University of Nebraska Press, 1999), 2:29. At the time Oviedo wrote the work, he had not yet traveled to parts of the mainland beyond Castilla del Oro. For this reason, his references to *tierra firme* speak particularly to his experiences in Castilla del Oro. He clarifies this position briefly in his account of "the things on *tierra firme*," writing, "I will say here something about what has been seen in these other provinces, or at least in those pertaining to the governance of Castilla del Oro" (*Dela natural hystoria*, 19).

19. Many chapters dedicated to *tierra firme* also mention the islands as well, albeit the former continues to occupy the center of attention. In the case of the chapter dedicated to pearl fishing, however, Oviedo focuses primarily on this activity as practiced on the islands, and in particular on Cubagua.

20. Merrim, "Apprehension," 170.

21. Carrillo Castillo, *Naturaleza*, 99.

22. Enrique Álvarez López ("La historia natural") provides the contemporary nomenclature for many species in the work and suggests possible modern-day equivalents for those that have not yet been successfully identified. Given that the aims of my research are not scientific, I will use the names that Oviedo provides for these species in their original Spanish or translated into English. Enrique Álvarez López, introduction and notes to *De la natural historia de las Indias*, by Gonzalo Fernández de Oviedo (Madrid: Editorial Summar, [1526] 1942).

23. *Dela natural hystoria*: "dígase de los otros mantenimientos que en la dicha isla usaban con que se sostenían de más de las frutas y pescados que esto está remetido adelante por ser común en todas las indias" (14 [chap. 6]).

24. *Dela natural hystoria*: "E tan poco en lo que toca a las frutas naturales de la tierra y a otras plantas y yeruas y a los pescados de mar y de agua dulce no curaré de ponerlo aquí en esta relación de la Española porque todo lo ay en la tierra firme más copiosamente y otras muchas más cosas que adelante en su lugar se dirán" (15 [chap. 7]).

25. David Watts, *The West Indies: Patterns of Development, Culture and Environmental Change Since 1492* (Cambridge: Cambridge University Press, [1987] 1994), 27.

26. Salas, *Tres cronistas*, 156–157.

27. María del Carmen Mena García, *Sevilla y las flotas de Indias: La gran armada de Castilla del Oro (1513–1514)* (Seville: Universidad de Sevilla, 1998), 185.

28. *Dela natural hystoria*, 32 [chap. 10].

29. *Historia general*, ed. Tudela Bueso, 3:252 [bk. 29, chap. 11].

30. Oviedo was on the island in 1515 and 1520, on his way to and from Spain, and again in 1523. It is unclear how much time he spent on the island in 1515, but he would not have wanted to lengthen his stay any more than necessary since he was eager to lodge his complaints against Pedrarias in Spain. Upon his return, in 1520, Oviedo stayed with his wife and children in Santo Domingo for eight days before they proceeded to Darién. Finally, Oviedo recalls recovering from an illness in Santo Domingo, where he remained for fifteen to twenty days before embarking for Spain in 1523. *Historia general*, ed. Tudela Bueso, 3:252 [bk. 29, chap. 11] and 3:285 [bk. 29, chap. 20].

31. Regarding his trip to Santo Domingo in 1523, Oviedo writes, "Vine al puerto de la Yaguana, en esta isla Española, desde a ocho días que partí de Cuba, e desde la Yaguana vine por tierra ochenta lenguas hasta aquesta ciudad de Santo Domingo" (*Historia general*, ed. Tudela Bueso, 3:285 [bk. 29, chap. 20]). It is likely that the sea and land route Oviedo describes here is the same as the one he took in 1515.

32. Fernández de Oviedo, *Memorias*, 1:311; *Historia general*, ed. Tudela Bueso, 3:303 [bk. 29, chap. 22].

33. *Historia* (1535): "otros escritores" (iii r).

34. *Dela natural hystoria*, 4 [prologue].

35. *Dela natural hystoria*: "que aunque acá se aya escrito y testigos de vista lo ayan dicho no será tan apuntadamente en todas estas cosas como aquí se dirá; aunque en algunas dellas o en todas ayan hablado la verdad los que a estas partes vienen a negociar o entender en otras cosas que de más interesse les pueden ser, los que les quitan de la memoria las cosas de esta calidad porque con menos atención las miran y consideran que el que por natural inclinación como yo ha desseado saberlas y por la obra ha puesto los ojos en ellas" (4 [prologue]).

36. Carrillo Castillo, *Naturaleza*, 36.

37. Paden, "Iguana," 206.

38. Merrim, "Apprehension," 191.

39. Gerbi, *Nature*, 362.

40. Sánchez Jiménez, "Memoria y utilidad," 271; Demetrio Ramos Pérez, "Las ideas de Fernández de Oviedo sobre la técnica de colonización en América," *Cuadernos hispanoamericanos* 17 (1957): 288.

41. *Historia general*, ed. Tudela Bueso, 3:62 [bk. 26, chap. 1].

42. *Historia general*: "para ser gobernada e poblada la tierra, y en más breve tiempo" (ed. Tudela Bueso, 3:62 [bk. 26, chap. 1]).

43. Barrera-Osorio, *Experiencing*, 4–5.

44. Barrera-Osorio, *Experiencing*, 102.

45. Carrillo Castillo, *Naturaleza*, 36; Paden, "Iguana," 206.

46. Merrim, "Apprehension," 190.

47. Frank Moya Pons, "The Politics of Forced Indian Labour in La Española, 1493–1520," *Antiquity* 66 (1992): 133–134.

48. Sauer, *Early Spanish Main*, 199.

49. Frank Moya Pons, *La Española en el siglo XVI, 1493–1520: Trabajo, sociedad y política en la economía del Oro* (Santiago, Dominican Republic: Universidad Católica Madre y Maestra Santiago, 1973), 173.

50. Sauer, *Early Spanish Main*, 200.

51. *Historia general*, ed. Tudela Bueso, 1:83–84 [bk. 3, chap. 12].

52. Watts, *West Indies*, 118.

53. Zuarez de Deza, "Relación de una carta a Su Alteza" (July 16, 1515): "Disce como los yndios van en muncha dysminución; e que convenia que Vuestra Alteza mandase a quien estaban encomendados, que dentro de cierto termino se casasen e que fysciesen vezindad por algund tiempo lymitado, e fuesen obligados a resydir en sus vezindades. . . . Que tambien se debe mandar a los que thobiesen yndios, que plantasen e sembrasen para probar la tierra, e se llevasen semillas para plantar huertas. . . . Disce que deberán animar a los Ofyciales de Vuestra Alteza a los casados, para que fysciesen ynxenios de azucar, que allá serían muy provechosos para aquellas partes" (*Colección de documentos inéditos* [*CDI*], 36:369–370).

54. *Cañafístola* is a Southeast Asian tree that produces pods that were used to purge and as a cure for several skin diseases. Watts, *West Indies*, 114. Oficiales de la Española, "Relación de una carta" (October 6, 1515): "Discen como rescebieron una carta de Vuestra Alteza, por la cual les disce que se maravilla del poco oro que abian recoxido, e que non es maravilla, por la dysminucion de los indios, e porque los que de antes los abian thenido, los abian muncho trabaxado. . . . Discen como la pez de Copey probaron en una nao ques ida a la Costa de las Peras. . . . Discen como an fecho buscar los pinos para bergantines, e que non an podido fallar nendguno; de otra madera tan Buena que se disce caoba, an fecho al presente fascer dos bergantines. . . . Discen como ymbian la caña fistula de la Ysla, e tambien ymbiaron la de la Ysla de Cuba, para que de la una e de la otra, Vuestra Alteza mande fascer espyriencia" (*CDI*, 36:426–427).

55. A *contador del rey* is a royal accountant. Real Academia Española, *Diccionario panhispánico del español jurídico*, s.v. "contador," https://dpej.rae.es/lema/contador1, accessed August 25, 2022.

56. Oficiales de la Española, "Relación de una carta" (August 7, 1515): "E que por relación de Gonzalo Fernandez de Oviedo, que viene por Procurador de la dicha Tierra-firme, podrá ser Su Alteza informado de como hay abundancia de manthenymientos en la dicha Tierra-firme . . . de que ynformará a Vuestra Alteza Xil Gonzalez Dávila, Su Conthador" (*CDI*, 36:381–382).

57. Although a footnote to the transcription of these memorials in the *Colección de documentos inéditos al descubrimiento [. . .]* dates them tentatively as having been written in 1518, it is more likely that they are from 1515 or 1516. *CDI*, 1:332; Manuel Giménez Fernández, *Bartolomé de las Casas: Delegado de Ciscernos para la reformación de las Indias (1516–1517)* (Seville: La escuela de estudios hispano-americanos de Sevilla, 1953), 450; Moya Pons, *La Española*, 183.

58. "Relación de Gil González Dávila": "Cuatro cosas principales han seydo causa de la disminución que digo en aquella Isla. La primera, la mudanza de los gobernadores. . . . La segunda causa ha seydo el mudar de los indios de unos vezinos en otros. . . . La tercera razón, que ha ayudado á esta otra, es los pelitos de entre los vecinos" (*CDI*, 1:332–333).

59. "Relación de Gil González Dávila": "La cuarta razón, e más principal ha seydo, que se ha tenido siempre fin á que de aquella Isla venga siempre el más oro que se pueda" (*CDI*, 1:333).

60. Spain's dependence on mineral wealth created a model of colonization and conquest opposed to that of the English, Portuguese, French, and Dutch who, not finding mineral wealth in significant quantities, would turn to agriculture as the economic base of their pres-

ence in the New World. John R. Fisher, *The Economic Aspects of Spanish Imperialism in America, 1492–1810* (Liverpool: Liverpool University Press, 1997), 13.

61. "Relación de Gil González Dávila," *CDI*, 1:337 and 1:340. The cultivation of sugar was not a new phenomenon to the island; Christopher Columbus brought sugarcane with him on his second voyage (1493–1496). A sugar industry did not start developing on Hispaniola until the second decade of the sixteenth century, however, when sugar prices in Europe increased and residents on the island began to consider alternative sources of livelihood to gold mining. Moya Pons, *La Española*, 257–258.

62. Moya Pons, *La Española*, 195.

63. Bartolomé de las Casas, *Historia de las Indias*, ed. Juan Pérez de Tudela Bueso (Madrid: Edición Atlas, 1961), 2:371 [bk. 3, chap. 85]. Between 1516 and 1518, Las Casas composed a series of memorials to denounce the situation in the Indies and propose solutions. He suggests incentives and policies to encourage colonists to cultivate the land and to ensure the physical and spiritual well-being of the natives. *Opúsculos, cartas y memoriales*, ed. Juan Pérez de Tudela Bueso (Madrid: Edición Atlas, 1958), 5:5–39.

64. Moya Pons, *La Española*, 223. A *licenciado* is an individual with a university degree. A *juez de residencia*, or judge of residency, made inquiries into the conduct of any given office holder in the colonial administration, usually as part of the process of finalizing the individual's period in office, and reported their findings to the Royal Council. Real Academia Española, *Diccionario panhispánico*, s.v. "juez de residencia," https://dpej.rae.es/lema/juez-de-residencia, accessed August 25, 2022.

65. "Al Cardenal Ximenex de Cisneros, los priores de San Gerónimo" (1517): "De verdad no es tierra para dejarla perder. . . . Si en esto se pone recado, esta isla Española será la mejor y más rica tierra e de más rentas" (*CDI*, 1:281–283).

66. "Al Cardenal Ximenex de Cisneros, los priores de San Gerónimo" (1517): "é dado que en su principio esto sea dificultoso por no estar la tierra cultivada ni tenerse noticia de los tiempos en que las simientes se deban sembrar, é las otras plantas desa tierra plantar; pero poniendo diligencia, mucho se podrá alcanzar con el ayuda del Señor" (*CDI*, 1:283).

67. "Capítulos de carta del licenciado Alonso de Çuaço al Emperador" (1518): "Tenga V.M. por cierto sea esta la mejor tierra del mundo" (*CDI*, 1:293).

68. Sauer, *Early Spanish Main*, 208–209; "Capítulos de carta del licenciado Alonso de Çuaço al Emperador" (1518), *CDI*, 1:293–295.

69. "Capítulos de carta del licenciado Alonso de Çuaço al Emperador" (1518): "En todo se entenderá con mucha diligencia" (*CDI*, 1:295).

70. *Historia general*, ed. Tudela Bueso, 3:251 [bk. 29, chap. 11].

71. *Historia* (1535), xxviii v [bk. 4, chap. 3].

72. *Historia* (1535): "porque el mudar la costumbre y especial en los indios es cortarles la vida" (xxxviii v [bk. 4, chap. 2]).

73. Moya Pons, *La Española*, 241.

74. Frank Moya Pons, "The Establishment of Primary Centres and Primary Plantations," in *General History of the Caribbean*, ed. Pieter C. Emmer (London: UNESCO, 1999), 2:68. The Crown encouraged the development of the sugar industry by providing loans to residents and sending technicians skilled in the construction of sugar mills from the Canary Islands to the Caribbean. Moya Pons, *La Española*, 263.

75. *Dela natural hystoria*: "Aquesta ysla fue muy poblada de indios . . . Que los indios que al presente ay son pocos y los christianos no son tantos quantos auer haber por causa que

muchos de los que en aquella ysla auia se han pasado a las otras yslas y tierra firme porque de más de ser los hombres amigos de novedades los que a aquellas partes van . . . porque como se han descubierto y descubren cada día otras tierras nuevas paresceles que en las otras hinchirán más ayna la bolsa" (6–7 [chap. 2]). In the *Sumario*, Oviedo does not query the cause and consequences of the demographic collapse of Hispaniola's native populations. Just as he understood New World flora and fauna to be the work of God, so too did he believe that native suffering was a divine response for them having lost or "forgotten" their ancestral knowledge of true religion. That is not to say that he excused all forms of Spanish violence against the indigenous peoples: in the *Historia general* Oviedo harshly criticizes those conquistadors who are known for their excesses and continues to advocate for a more peaceful model of conquest and colonization.

76. *Dela natural hystoria*, 7 [chap. 2].

77. *Dela natural hystoria*: "Si un príncipe no touviesse más señorío de aquesta ysla sola en breue tiempo sería tal que ni le haría ventaja Secilia ni Inglaterra ni al presente ay de que pueda tener embidia a ninguna delas que es dicho: antes lo que enla ysla Española sobra podría hazer ricas a muchas provincias y reinos" (7 [chap. 2]).

78. *Dela natural hystoria*: "más ricas minas y de mejor oro que hasta oy en parte del mundo en tanta cantidad se ha hallado ni descubierto" (7 [chap. 2]).

79. *Dela natural hystoria*, 7–8 [chap. 2]. *Dela natural hystoria*: "Se hazen muy mejor y en más cantidad que en parte de nuestra Europa" (7 [chap. 2]).

80. *Dela natural hystoria*, 8 [chap. 2].

81. *Dela natural hystoria*: "Las naos vienen cargadas della cada vn año" (7 [chap. 2]).

82. *Dela natural hystoria*: "Ya se trae a España en mucha cantidad: y desde ella se lleua y reparte por muchas partes del mundo" (7 [chap. 2]).

83. *Dela natural hystoria*, 7 [chap. 2].

84. *Dela natural hystoria*: "Desta causa no se dan a hacer pan ni a poner viñas, porque en aquel tiempo que estas cosas tardarán en dar fruto, las hallan en buenos precios y se las llevan las naos desde España, y labrando minas o exercitandose en la mercadería o en pesquerías de perlas o en otros exercicios como he dicho más presto allegan hacienda" (7 [chap. 2]).

85. *Dela natural hystoria*, 7 [chap. 2].

86. *Historia* (1535), xxxii r–xxxii v [bk. 3, chap. 11].

87. Fisher, *Economic Aspects*, 26–27.

88. *Historia* (1535): "Qual ingenio mortal sabrá comprehender tanta diversidad de lenguas, de habito, de costumbres en los hombres destas indias. Tanta variedad de animales assi domésticos como salvajes y fieros. Tanta multitude innarrable de árboles copiosos de diversos géneros de frutas y otros estériles. . . . Quantas plantas y yerbas útiles y provechosas al hombre. Quantas otras innumerables que a él no son conoscidas y con tantas diferencias de rosas y flores y olorosa fragancia" (ii v [bk. 1], translation based on Merrim, "Apprehension," 168–169).

89. *Historia* (1535): "Las quales ha rescebido esta tierra no como madrastra sino como más verdadera madre que la que selas embió, pues en más cantidad y mejor que en España se hazen algunas dellas, assí de los ganados vtiles . . . como de pan y legumbres y frutas y açucar" (ii v [bk. 1]). Translation based on John Murray, review of *Chronological History of the West Indies*, by Capt. Thomas Southey, *Quarterly Review* 38 (1828): 197.

90. *Historia* (1535), ii r [bk. 1].

91. *Historia* (1535): "E assi lo podrían hazer de otras cosas que acá están oluidadas y aquestas indias antes que los españoles las hallassen produzian y agora produzen" (ii r [bk. 1]).

92. *Historia* (1535), ii r [bk. 1]. Orchella weeds are species of lichen from which are extracted a purple dye, orcein. Prior to Columbus's voyages to the New World, the Spanish Crown collected it from the Canary Islands. Liss, *Isabel*, 161.

93. In a similar vein, Louise Bénat-Tachot argues that "experience" acquires diverse meanings in the *Historia general*. If initially Oviedo touts its importance for accurately depicting man and nature, soon he comes to see first-person knowledge of New World geography and indigenous customs as necessary for the success of the colonial enterprise in the Americas. Bénat-Tachot, "Experiencia," 257.

94. *Historia* (1535): "Ninguna manera de manjar hallamos en estas partes que fuesse como aquel que nos dieron nuestros padres, el pan de rayzes, las frutas salvajes y no conocidas ni conformes a nuestros estómagos, las aguas de diferentes gustos las carnes ningunas se hallaron en esta isla sino aquellos gozques mudos que he dicho y otros pocos animales y muy diferentes a los de España. Sierpes que son las yvanas, culebras y lagartijas. . . . Y como todas estas cosas auia provado este capitán enla isla de Sant Juan y enla tierra del norte en la Florida y otras partes supo dares maña para lo que está dicho y sin duda yo creo que si a ello fuera vno que de España viniera nuevamente nunca la paz se concluyera" (lv r–lv v [bk. 5, chap. 9]).

Chapter 4

1. The 1535 edition of the *Historia* comprises part 1, or books 1 to 19, of what would eventually become a three-part work divided into fifty books, as well as an additional book on shipwrecks. The latter, which Oviedo appended to the *Historia* (1535) as *Book 20*, would become *Book 50* in the Monserrat manuscript.

2. *Historia* (1535): "Digo que en general los árboles que en estas Indias ay es cosa para no se poder explicar por su multitud porque la tierra está tan cubierta de los en la mayor parte y con tantas diferencias y desemejança los vnos delos otros assi en la grandeza como en el tronco y las ramas y en la hoja y en el aspeto y en la fruta que ni los indios naturales los conocen ni saben dar nombres por su multitud . . . y en muchas partes no se puede ver el cielo desde debajo destas arboledas por ser tan altas y tan espessas y llenas de rama, y en muchas partes no se puede andar entre ellas, porque . . . ay otras plantas y verduras tan texidas y rebueltas y de tantos espinos y bexucos y otras ramas mezcladas que con mucho trabajo y afuerça de puñales y hachas es menester abrir el camino. Y lo que en esto se podría decir es un mare magno y oculto: porque aunque se vee lo más dello se ygnora: porque no se saben como he dicho los nombres a tales árboles ni sus propiedades. . . . Los vnos tienen fruta y los otros están en flora: otros comiençan a brotar. Y assi como son de diversos géneros assi gozan del tiempo en diferente manera. Y se vee todo junto en una sazón y en cualquier parte del año: y por tanto dexaré aquesto porque desta infinidad de géneros y multitud de diferencias con el tiempo se yrán entendiendo muchas cosas que al presente no se saben" (lxxxvii r–lxxxvii v [bk. 9, prologue]).

3. Gerbi, *Nature*, 131; Bleichmar, *Visual Voyages*, 20; Portuondo, "Cosmography," 68.

4. The second edition, titled *Coronica de las Indias*, was published in Salamanca in 1547 by Juan de Junta. There is some dispute among scholars as to whether Oviedo authorized this publication, since he does not make any references to it in the Monserrat manuscript, on which he was working at the time. Nevertheless, circumstances seem to suggest that Oviedo had a hand in its reprint: he was in Salamanca in the spring of 1547, the place and date of the publication of the second edition. Although Ramos suggests that the *Historia* (1547) expands on the first edition (as its title page does indeed claim), Turner's exhaustive analysis of the two

reveals that they are virtually identical, "save for slight variations in abbreviations and punctuation" (Turner, "Aborted," 109). Demetrio Ramos Pérez, "Fernández de Oviedo y el 'enigma' de la edición de 1547 de su *Historia general*," *Revista historia* 1, no. 8 (1969): 451.

5. *Dela natural hystoria*: "general y natural historia de Indias" (103).

6. *Historia* (1535), bk. 1.

7. *Historia* (1535): "mejor y más copiosamente dicho" (iii v [bk. 1])

8. *Historia* (1535), iii r [bk. 1].

9. Carrillo Castillo, *Naturaleza*, 116. It is possible that Oviedo continued to add to the *Historia general* after having left the Monserrat manuscript in Spain and returning to Santo Domingo. A passage in *Quinquagenas de la nobleza de España* supports this theory; according to the epistle that Oviedo includes in the work's prologue, dedicated to Prince Philip and dated January 10, 1555, the three-part, fifty-book *Historia general* was at that moment being printed, while Oviedo was writing a fourth part. This volume has never been located, but there is no reason to doubt Oviedo's claims; a constant work in progress, Oviedo describes the project in the letter to the prince as "a never-ending history, longer than my life" (*Memorias*, 1:19).

10. Although Oviedo states in 1555 that the three-part, fifty-book *Historia general* was at that moment being printed, no such publication exists. Daymond Turner dates the Ayer copy of Oviedo's *Book 20* to that same year, proposing that this manuscript is evidence that Oviedo had attempted to publish at least nine books of the second part of the *Historia general* around that time. Lack of funds put the printing temporarily on hold, and Oviedo's death in 1557 brought it to its definitive conclusion. Turner, "Aborted," 118–119.

11. In point of fact, there appears to have been more variety in island flora than what Oviedo acknowledges or was aware of. On Cuba, forty-one plant genera, or roughly 50 percent of its species, are endemic to the island, meaning they are found nowhere else. On Hispaniola, this number is twenty-six (33 percent), and 27 percent of Jamaica's angiosperm species (plants that produce flowers) are endemic. Watts, *West Indies*, 25.

12. *Historia* (1535): "porque en muchas cosas tienen semejança assi enlas aves como en los animales y en las pesquerías y otras particularidades" (cxvii r [bk. 16, prologue]).

13. *Historia* (1535), cxxxii r–cxxxii v [bk. 17, chap. 4].

14. Descriptions of these phenomena are found in the *Historia* (1535) on the following pages: *palo santo* (cxxviii r [bk. 16, chap. 17]), the oil well in Cuba (cxxxiii v [bk. 17, chap. 8]), and fishing in Jamaica (clii r [bk. 18, chap. 2]).

15. Álvarez López, introduction and notes to *De la natural historia de las Indias* (Madrid: Editorial Summar, [1526] 1942), 199n6; Watts, *West Indies*, 37. Although much of the island fauna, like its flora, originated from the South American continent, its variety and diversity were limited by the fact that many mainland animal species could not travel easily to the islands. As the geographer David Watts explains, "because of the difficulties experienced by some species, genera and families in crossing the intervening water bodies, the basic faunal structure of much of the region is peculiarly imbalanced, and, in the case of every island, impoverished when compared to equivalent areas of the South American continental mainland" (*West Indies*, 37).

16. Oviedo was uncertain whether rats had originated on the island, and he searched in vain for a witness who could recall if they had been present there upon Columbus's arrival. *Historia* [1535], xcix v [bk. 12, chap. 6]. At least three species of rat, the black rat (*Rattus rattus*), the brown rat (*R. norvegicus*), and the roof rat (*R. rattus alexandrinus*) were introduced

from Europe. Native fauna disappeared from areas near larger human settlements on Hispaniola, such as harbor towns and *villas*. Such is the case with the hutia and the iguana. Still others were killed by species imported from Europe—namely, the feral cat and several species of rat. Watts, *West Indies*, 120.

17. *Historia* (1535): "según muchos afirman que es de quatro pies y tan grande como vn podenco o sabueso mediano, y es de color pardo como la hutia y del mismo talle o manera salvo quel quemi es mucho mayor" (xcviii v [bk. 12, chap. 2]).

18. *Historia* (1535): "Todo lo que fuesse de comer les parecería entonces muy bueno y sabroso aun que no lo fuesse" (xcviii v–xcix r [bk. 12, chap. 2]).

19. A key difference between the *Sumario* and the *Historia naturalis* is that, for Pliny, animals take precedence over all plants, whereas agriculture is the first subject Oviedo addresses with regard to New World nature in both the *Sumario* and the *Historia general*.

20. It is unclear when Oviedo settled on this new organization. There is evidence to suggest that he continued to envision the *Historia* (1535) as replicating the order of the *Sumario* even after he had begun the project. In the prologue to *Book 7: On Agriculture*, the first book in the *Historia* (1535) dedicated to New World nature, Oviedo indicates that he will address these phenomena in the order he had in the *Sumario*: land and water animals, birds, insects, trees, plants, and herbs. In the end, however, Oviedo decided to place the books on island flora first. Oviedo, *Historia* (1535), lxxi v [bk. 7, prologue].

21. Carrillo Castillo, *Naturaleza*, 101.

22. *Historia general*, ed. Tudela Bueso, 1:226 [bk. 7, chap. 1].

23. Rolena Adorno, *Colonial Latin American Literature: A Very Short Introduction* (Oxford: Oxford University Press, 2011), 24.

24. *Historia* (1535): "para todos lo que viven, y después del nos vinieren" (ii v [bk. 2, chap. 2]).

25. *Historia* (1535), iii r [bk. 2, chap. 3].

26. *Historia* (1535), ii r [bk. 2, chap. 1], iii r [bk. 2, chap. 3], iiii v [bk. 2, chap. 3].

27. Álvaro Félix Bolaños explores the Hesperides thesis more extensively in "The Historian and the Hesperides: Fernández de Oviedo and the Limitations of Imitation" (*Bulletin of Hispanic Studies* 72 [1995]: 273–288). Oviedo's argument is complex and convoluted; Rolena Adorno summarizes it most effectively (*Colonial*).

28. *Historia* (1535), iii r [bk. 2, chap. 3].

29. *Historia* (1535), cxvii r–cxvii v [bk. 16, prologue].

30. *Historia* (1535): "torno a dezir lo que he dicho y escripto otras vezes y es: que si vn príncipe no tuviesse mas señorio de aquesta ysla sola en breue tiempo seria tal que fiziesse ventaja a las yslas de secilia y inglaterra porque lo que aquí sobra a otras prouincias haria muy ricas" (xxxi v [bk. 3, chap. 11]).

31. Adorno, *Colonial*, 25; Bolaños, "Historian," 278.

32. Adorno, *Colonial*, 25.

33. Oviedo's Hesperides thesis provoked outrage and concern among Columbus's heirs, who were in the midst of ligation (1511–1538) to defend the privileges that the Crown had originally granted to the explorer. Yet Oviedo's intention was never to deny Columbus his due but rather to highlight his great service to the Crown for having "rediscovered" the Indies. Carrillo Castillo, introduction to *Oviedo on Columbus*, ed. Jesús Carrillo Castillo, trans. Diane Avalle-Arce (Turnhout, Belgium: Brepols, 2000), 18–19; Marcel Bataillon and Edmundo O'Gorman, *Dos concepciones de la tarea histórica con motivo de la idea del descubrimiento de América* (Mexico City: Imprenta Universitaria, 1955), 34. In arguing that Spain

had possessed these lands in ancient times, Oviedo is much less concerned with these legal battles than he is with establishing the "dynastic rights [of Spain] over the Antilles prior to the 1492 Capitulations of Santa Fe" (Carrillo Castillo, introduction, 18–19).

34. Oviedo does not hypothesize about the origins of the inhabitants of the Indies, as does the sixteenth-century Jesuit missionary and naturalist José de Acosta (1540–1600). In Oviedo's account, the uninhabited islands discovered by the Carthaginians inexplicably became peopled prior to the arrival of the apostles, several centuries later. *Historia* 1535, ix r–ix v [bk. 2, chap. 7].

35. *Historia* (1535), cxvii r [bk. 16, prologue].

36. *Historia* (1535), cxvii r [bk. 16, prologue].

37. Adorno, *Colonial*, 25. *Historia* (1535): "La otra parte o mitad del mundo es aquesta de nuestras Indias" (cxvii r [bk. 16, prologue]).

38. *Historia* (1535): "La santa iglesia ha ya predicado en todas las partes del mundo el misterio de nuestra redempción" (ix v [bk. 2, chap. 7]).

39. *Historia* (1535), ix v [bk. 2, chap. 7].

40. *Historia*: "paresce el testimonio de la Cruz, e que no la han podido ignorer estas gentes" (ed. Tudela Bueso, 1:252 [bk. 8, chap. 4]).

41. Wey Gómez, *Tropics*, 74.

42. Nicolás Wey Gómez, "Memorias de la zona tórrida: El naturalismo clásico y la 'tropicalidad' Americana en el *Sumario del natural historia de las Indias* de Gonzalo Fernández de Oviedo (1526)," *Revista de Indias* 259 (2013): 618.

43. *Dela natural hystoria*, 24 [chap. 10].

44. Wey Gómez, *Tropics*, 269.

45. *Dela natural hystoria*: "Verdad es que, según las marauillas del mundo y los estremos que las criaturas, más en vnas partes que en otras, tiene, según las diuersidades delas provincias y constelaciones donde se crían, ya vemos que las plantas que son nociuas en vnas partes, son sanas y provechosas en otras. Y las aves que en vna provincia son de buen sabor, en otras partes no curan dellas ni las comen" (37 [chap. 11], trans. Barrera-Osorio, *Experiencing*, 111).

46. *Historia* (1535), xc v–xci r [bk. 9, chap. 16] and clix r [bk. 19, chap. 8].

47. Wey Gómez, "Memorias," 611–612.

48. *Historia* (1535): "mitad del mundo (cxvii r [bk. 16, prologue).

49. *Historia* (1535): "Quantos montes mas admirables y espantosos que ethna" (ii v [bk. 1]).

50. *Historia* (1535): "tan grandes reynos y provincias y de tan extrañas gentes y diuersidades y costumbres y cerimonias y ydolatrias apartadas de quanto esaua escripto (desde ab inicio hasta nuestro tiempo)" (ii v [bk. 1]).

51. *Historia* (1535): "Pues no te puedo comprehender, comprehendeme tú a mí" (xv v [bk. 2, chap. 10]).

52. *Historia* (1535), xv v–xvi r [bk. 2, chap. 10].

53. Augustine, *Confessions*, ed. and trans. Carolyn J. B. Hammond (Cambridge, MA: Harvard University Press, 2016), bk. 10, sec. 55.

54. Augustine, *Confessions*, bk. 10, sec. 55.

55. *Historia* (1535): "de que proceden los efectos delas nouedades que recuento, porque ni soy tan philosopho para comprehenderlos ni me quiero detener en argumentos, sino conforme a la vista diré lo que he podido comprehender o he sentido en estas materias" (cxii v [bk. 15, chap. 1]).

56. *Historia* (1535): "Yo los he mirado y parecenme muy diferentes estos de aca de lo que él [Plinio] escriue" (lxxxix r [bk. 9, chap. 10]).

57. *Historia* (1535): "Y el que concierta en vna señal se desacuerda en otras, y assi por una sola que veen los que no tienen esperiencia en las cosas le conceden el nombre assi como si tuviesse todas las partes y circunstancias que Plinio diz" (lxxxix v [bk. 9, chap. 10]).

58. *Historia* (1535): "Es cierto que yo he muchas vezes ocupado me inquiriendo este árbol por mi persona hallandome por estos caminos y boscajes y en diversas partes de estas yslas y tierra firme destas indias" (lxxxix r–lxxxix v [bk. 9, chap. 10]).

59. Oviedo questions the names given to other plants in the Indies—for example, the so-called cedar and the taray—but not nearly with the level of attention and concern that he devotes to the case of the terebinth. Daniel Egaña Rojas, "Comerse las Indias: La alimentación como clave clasificatoria del Nuevo Mundo en la obra de Fernández de Oviedo," *Anunario de Estudios Americanos* 72, no. 2 (2015): 583.

60. Ogilvie, *Science*, 185.

61. The rigor and meticulousness that characterizes Oviedo's approximations to New World nature is at odds with Carrillo Castillo's efforts to locate the main source of his knowledge in "the spontaneous understanding of things derived from direct experience" ("The Eyes of the New Pliny: The Use of Images in Gonzalo Fernández de Oviedo's *Historia general y natural de las Indias*," in *The Art of Natural History: Illustrated Treatises and Botanical Paintings, 1400–1850*, ed. Therese O'Malley and Amy R. W. Meyers [Washington, DC: National Gallery of Art, 2008], 112).

62. This is not the only time that Oviedo takes issue with the arbitrary naming of phenomena in the Indies, ranging from plants and animals to places. In criticizing instances of erroneous or inappropriate labeling, Carrillo Castillo argues that Oviedo not only undermines others' authority and expertise but also seeks to define "the New World as an autonomous and self-contained entity, bearing its own labels and demarcations" ("Naming Difference: The Politics of Naming in Fernández de Oviedo's *Historia general y natural de las Indias*," *Science in Context* 16, no. 4 [December 2003], 497).

63. *Historia* (1535): "algo más altas que vn hombre y otras mucho menores gruessas como dos dedos y algunas más y otras menos" (lxxii v [bk. 7, chap. 2]).

64. *Historia* (1535), lxxii r [bk. 7, chap. 1].

65. *Historia* (1535): "tenia más de veinte pies de luengo y lo más gruesso della era mucho más que vn puño cerrado" (c r [bk. 12, chap. 8]).

66. *Historia* (1535): "Porque muchos hombres . . . estendidos los braços prouaron si alcançarían con su braza de punta a punta delas alas deste rabiohorcado que he dicho teniendo las abiertas y tendidas, y con más de quatro dedos ninguno alcançó" (cix r [bk. 14, chap. 1]).

67. *Historia* (1535): "que no pudiera yo creerlo que allí vi por esperiencia a ninguno que tal me dixera" (cix r [bk. 14, chap. 1]).

68. Cañizares-Esguerra, *Nature, Empire, and Nation*, 120.

69. Prieto, *Missionary*, 206.

70. *Historia* (1535): "No ay cosa errada ni mal conpuesta enla natura, por qual maestro y hazedor della no pudo errar ni hizo cosa desconuiniente ni sin prouecho" (cliii r [bk. 19, prologue]).

71. *Historia* (1535): "Son vna manera de cardos muy espinosos y saluajes que no ay en ellos parte de donde se puedan tocar sin muy fieras espinas" (lxxxv v [bk. 8, chap. 24]).

72. *Historia* (1535): "La natura selas pone por orden y atrechos vnas de otras con mucha orden y compas en su composición" (lxxxv v [bk. 8, chap. 24]).

73. *Historia* (1535), lxxii v [bk. 7, chap. 1].

74. Daymond Turner, "Gonzalo Fernández de Oviedo's *Historia General y Natural*: First American Encyclopedia," *Journal of Inter-American Studies* 6, no. 2 (April 1964): 273.

75. Oviedo recounts several culinary uses for corn that he learned from native informants. For example, he attributes the making of cornbread to indigenous communities on *tierra firme*. He also describes the "big, thin, and white" *tortas* made in Nicaragua that, according to Oviedo, have their origin in New Spain (*Historia* 1535, lxxii v [bk. 7, chap. 1]). He elaborates on the preparation of yuca as well, both as a bread and a kind of liquor. *Historia* 1535, lxxiii v–lxxiiii r [bk. 7, chap. 1].

76. *Historia* (1535), cix r [bk. 14, chap. 1]; Egaña Rojas, "Comerse," 585; Oviedo, *Historia* (1535): "se aprendió de los indios" (xciiii v [bk. 10, chap. 4]).

77. *Historia* (1535), lxxxi r [bk. 8, chap. 2] and cxvi v [bk. 15, chap. 8].

78. *Historia* (1535): "se aprendió de los indios y assi lo hazen ellos" (lxxii r [bk. 7, chap. 1]); *Historia* (1535), xcv r [bk. 10, chap. 15].

79. *Historia* (1535): "Y assi creo que fueron todas las del mundo en su principio, y delas tales se tomó el origen para las cultiuar y hazer mejores" (lxxix r [bk. 8, chap. 1]).

80. Ken Albala, *Eating Right in the Renaissance* (Berkeley: University of California Press, 2002), 123.

81. *Historia* (1535): "Y no ay duda sino que donde la natura y su oficio produze algunas destas cosas que mejor se harán ayudando en ello los hombres por el regar y otras diligencias que los hombres alcançan de los secretos dela agricultura assi como el enxerir, el podar, el estercolar, y regar a sus tiempos" (lxxxv r [bk. 8, chap. 21]).

82. *Historia* (1535): "Los cristianos hazen muy mejor a causa del arar la tierra en donde ay disposición para ello" (lxxii r [bk. 7, chap. 1]).

83. *Historia* (1535), lxxviii v [bk. 8, chap. 1] and lxxxii v [bk. 8, chap. 9].

84. *Historia* (1535), lxxxviii r [bk. 9, chap. 3].

85. *Historia* (1535), xlvi r [bk. 5, chap. 1].

86. *Historia* (1535): "Conoscen muchas yeruas y tienen las esperimentadas para esto y para otras muchas dolencias" (xciii r [bk. 10, chap. 2]).

87. *Historia* (1535): "Naturalmente pudieron estas gentes salvajes venir en conocimiento dellas" (xxv v [bk. 3, chap. 5]).

88. *Historia* (1535): "Esta ynorancia viene de no entender ellos que aquello es propiedad dela natura" (cvii v [bk. 13, chap. 10]).

89. *Dela natural hystoria*: "La cosa que más conserua y sostiene las obras de natura en la memoria delos mortales son las ystorias y libros en que se hallan escriptas: y aquellas por mas verdaderas y auténticas se estiman que por visto de ojos el comedido entendimiento del hombre que por el mundo ha andado se ocupo en escreuirlas" (3 [prologue]).

90. *Historia general*, ed. Tudela Bueso, 5:417 [bk. 50, chap. 30].

91. *Historia* (1535): "Pero como los indios antiguos son ya muertos assi se ha acabado con ellos el conoscimiento destas propiedades y secretos de natura. Digo delos que estaban por los indios esperimentados o sabidos: y todo lo que agora se puede dezir es poco y no bien entendido porque esta generación es tan auara desso poco que sabe que por ningún interesse ni bien que se les haga quiere manifestar cosa destas, en especial delas que podrían aprouechar a los cristianos si son medicinales" (xcvii v [bk. 11, chap. 5]).

92. Bénat-Tachot, "La experiencia," 256–257.

93. Prieto, *Missionary*, 42.

94. Bleichmar, *Visual Voyages*, 51; Francisco Hernández, *The Mexican Treasury: The Writings of Dr. Francisco Hernández*, ed. Simon Varey, trans. Rafael Chabrán, Cynthia L. Chamberlin, and Simon Varey (Stanford, CA: Stanford University Press, 2000), 254.

95. *Historia* (1535): "Y como esperimentado quiero avisar a quien esto leyere en estas partes (si indios faltaren como ya faltan) de la manera y arte que ha de tener para guisar los huesos de la yvana" (cii v [bk. 13, chap. 3]).

96. *Historia* (1535): "Estas esperiencias pocos indios las saben ya hazer: porque los viejos son muertos: y porque los cristianos no lo han menester porque para agro ay tantas naranjas y limones enla isla que no ay necessidad de lo que es dicho: ni para licor dulce mucho menos por auer tanto açucar en la ysla: y assi se ha oluidado lo que en estos dos casos de dulce y agro seruía el çumo de la yuca" (lxxiiii r [bk. 7, chap. 2]).

97. Fernández de Oviedo, *Historia* (1535), lxxxviii r [bk. 9, chap. 3]; ciii v [bk. 13, chap. 4]; xcv r [bk. 10, chap. 6].

98. Smith, *Body of the Artisan*, 66; Vives, *Vives*, 209.

Chapter 5

1. Oviedo's verbal and visual description of the pineapple in the *Historia general* has been analyzed extensively by such scholars as José Rabasa and Kathleen Ann Myers as epitomizing these epistemological difficulties. José Rabasa, *De la invención de América: La historiografía española y la formación del eurocentrismo*, trans. Aldo Mazzucchelli (Mexico City: Universidad Iberoamericana, [1993] 2009), 165–170; Myers, *Fernández*, 75–80.

2. *Historia* (1535): "No puede la lengua dar tan particularmente la razón y pintura de esta fruta que satisfaga del todo lo que se requiere particularizar en este caso. Y porque yo no sé también debuxar como quería para que demás de las palabras en alguna manera la vista del letor pudiesse participar de esta verdad pornéla como supiere hazer lo en la margen. Puesto que assi por no tener colores como porque le faltara mucho al debuxo no satisfaré tan copiosamente como la materia lo pide para dar la de todo punto a entender" (lxxvi v [bk. 7, chap. 13]).

3. These include, in order of appearance: a coat of arms (x r); a constellation (xvi v); a drum (xlvi v); a tobacco pipe (xlvii r); a hammock (xlvii v); two types of indigenous dwellings (lviii v, lix r); a canoe (lxi r); an ax (lxi r); sticks for sparking a fire (lxi v); the mining of gold (lxvi r); two types of yuca leaf (lxxiii r); a pineapple (lxxvi v); a plantain leaf (lxxix r); an *higüero* leaf (lxxxi r); a *guiabara* leaf (lxxxiii r); a copey leaf (lxxxiii v); a *guanábana* fruit (lxxxiiii r); a mamey leaf (lxxxiiii v); a pitaya plant (lxxxv r); a cactus (lxxxvi r); a tuna plant (lxxxvi v); a cactus Oviedo refers to as the tree that heals broken bones (*el árbol de las soldaduras*) (xcii r); an artificial balsam leaf (xcvii v); a *perebecenuc* leaf (xcviii r); an iguana (ciii v); a manatee (cvi r); and a shell (clix r).

4. The expense of producing woodcuts was not inconsequential at the time, providing Oviedo with an added incentive to ensure their accuracy. Clive Griffin, *The Crombergers of Seville: The History of a Printing and Merchant Dynasty* (Oxford: Clarendon Press, 1988), 185.

5. Griffin, *Crombergers*, 185–186.

6. Luis de Santángel was an *escribano de ración*, overseeing the royal household's finances and administration.

7. William Sturtevant, "The Sources for European Imagery of Native Americans," in *New World of Wonders: European Images of the Americas, 1492–1700* (Seattle: University of Washington Press, 1992), 25; and Bleichmar, *Visual Voyages*, 3.

8. Sturtevant, "Sources," 27.

9. The remainder of the woodcuts in the *Sumario* and the *Historia* (1535) depict indigenous artifacts and, to a lesser degree, animals.

10. Turner, "Forgotten Treasure," 35.

11. The Paraguayan Natalicio González (1897–1966) edited the *Historia general* (1944–1945) as well, reproducing Ríos's interventions. By far the most commonly cited modern edition of the *Historia general*, however, is that of Tudela Bueso.

12. Myers has drawn renewed attention to the original woodcuts and drawings of the *Historia general* by compiling an inventory of many of them in appendix E of *Fernández de Oviedo's Chronicle of America* (2007). Her efforts highlight the risks in dealing with a work with such a complicated print and manuscript history. Myers provides a table listing the location of each image (including the page and work from which it was taken). However, by neglecting to explain in detail the process by which Oviedo composed the work as well as its subsequent manuscript history, she renders the table useful only to those already well acquainted with this background. Moreover, portions of the table are misinformed: Myers groups under one column heading all the print editions of the *Historia* (1535, 1547, and 1557), overlooking the differences in their illustrated material as well the circumstances of their printing. In short, Myers's attempts to do justice to the complete corpus of Oviedo's visual production shed light on the care that one must take when dealing with such disparate materials.

13. Gerbi, *Nature*, 182; Álvarez López, "La historia natural," 591. Although Álvarez López describes the illustrations in the German botanist Leonhart Fuchs's *De historia stirpium commentarii insignes* (1542) as being beautiful and exact, it does not appear that he ascribes to them any scientific importance. He distinguishes between these woodcuts and the scientific merits of the work, which, he claims, in no way rival the contributions of Oviedo's writings on the Indies.

14. Turner, "Forgotten Treasure," 42.

15. The illustrations in Oviedo's writings in the Indies have been reproduced in numerous articles and monographs, ranging from those tracing the development of the genre of New World natural history and the fields of botany, zoology, and anthropology, to more focused studies on the author and his works. Very infrequently, however, do scholars analyze these illustrations extensively, as part of a coherent and ever-changing visual program. Scholars who have engaged with this topic recently include Carrillo Castillo, *Naturaleza*; Carrillo Castillo, "Eyes," 109–125; Myers, *Fernández*; Myers, "Representation," 183–213; and Vanina Teglia, "Conocer el Nuevo Mundo: Referencialidad en las ilustraciones de Fernández de Oviedo," *Olivar* 15, no. 2 (December 2014): 1–14.

16. Myers makes a similar observation, arguing that these illustrations are often studied simply for their documentary value as some of the earliest visual records of New World flora and fauna. *Fernández*, 64–65.

17. Carmen Niekrasz and Claudia Swan, "Art," in *The Cambridge History of Science: Early Modern Science*, ed. Katharine Park and Lorraine Daston (Cambridge: Cambridge University Press, 2006), 3:794.

18. Karen M. Reeds, "Leonardo da Vinci and Botanical Illustration: Nature Prints, Drawings, and Woodcuts ca. 1500," in *Visualizing Medieval Medicine and Natural History, 1200–1550*, ed. Jean A. Givens, Karen M. Reeds, and Alain Touwaide (Aldershot: Ashgate, 2006), 236.

19. Bleichmar, *Visible Empire*, 6.

20. Bleichmar, *Visible Empire*, 5–6.

21. Wilfrid Blunt and Sandra Raphael, *The Illustrated Herbal* (New York: Thames and Hudson, [1979] 1994), 15–16. This codex did not resurface until 1406, in Constantinople. After conquering the city in 1453, the Turks took possession of the manuscript.

22. Blunt and Raphael, *Illustrated*, 113.

23. Jerry Stannard, *Herbs and Herbalism in the Middle Ages and Renaissance*, ed. Katherine E. Stannard and Richard Kay (Farnham: Ashgate, 1999), 31.

24. Brian W. Ogilvie, "Image and Text in Natural History, 1500–1700," in *The Power of Images in Early Modern Science*, ed. Wolfgang Lefèvre, Jürgen Renn, and Urs Schoepflin (Basel: Birkhäuser, 2003), 143.

25. Sachiko Kusukawa, "Leonhart Fuchs on the Importance of Pictures," *Journal of the History of Ideas* 58, no. 3 (1997): 416.

26. Ogilvie, *Science*, 196.

27. Kusukawa, "Leonhart Fuchs," 411.

28. Ogilvie, *Science*, 197.

29. Pliny, *Naturalis historia*, bk. 25, chaps. 4–5.

30. *Dela natural hystoria*: "Y para darse a entender sería necessario pintarse para que por medio dela vista se comprehendiesse lo que la lengua falta en esta parte" (84 [chap. 80]).

31. *Historia*: "Sin dubda, los ojos son mucha parte de la información destas cosas, e ya que las mismas no se puedan ver ni palpar, mucha ayuda es a la pluma la imagen dellas" (ed. Tudela Bueso, 1:268 [bk. 8, chap. 30]).

32. Carrillo Castillo, *Naturaleza*, 297–299; Myers, *Fernández*, 77.

33. *Dela natural hystoria*, 78 [chap. 78].

34. *Dela natural hystoria*, 79 [chap. 78].

35. *Historia general*: "He visto otros muchos y muy mayores" (ed. Tudela Bueso, 1:290 [bk. 9, chap. 11]).

36. In later stages of his project, however, Oviedo does attempt to illustrate other natural phenomena even bigger than the trivet tree: volcanoes. The first of these images, in *Book 33* of the Monserrat manuscript, demonstrates his inability to render topographical features successfully: so small is the volcano and disproportionate its parts that it is necessary to consult the written description to know what is being represented. In contrast, the three drawings of volcanoes copied by Andrés and Antonio Gascó in the 1560s from the now-lost *Book 42* of the Monserrat manuscript are more easily recognizable; they are larger, hand-colored, and include details taken from the surrounding landscape. Unfortunately, it is impossible to ascertain the extent to which these images are faithful to the originals. Gonzalo Fernández de Oviedo, *Adición al libro 14 de la segunda parte*, 9/6480 (14), Biblioteca de la Real Academia de la Historia, Madrid, Spain, 156r [bk. 33]. Gonzalo Fernández de Oviedo, *Tercera parte de la Historia natural y general de las Indias*, II/3042, Biblioteca del Palacio Real, Madrid, Spain, 60v–62r [bk. 42]. Carrillo Castillo has written more extensively about Oviedo's account of the Masaya volcano: "From Mt Ventoux to Mt Masaya: The Rise and Fall of Subjectivity in Early Modern Travel Narrative," in *Voyages and Visions: Toward a Cultural History of Travel*, ed. Jás Elsner and Joan-Paul Rubiés (London: Reaktion Books, 1999), 57–73.

37. In equatorial South America the general division between plantain and banana would have been largely redundant, given that both were used as a vegetable when green and a fruit when ripe. In the *Historia* (1535), Oviedo recounts that the Dominican friar Tomás de Berlanga brought plantains to Hispaniola from the Canary Islands in 1516. It is possible that the cultivars Berlanga transplanted did not originate from just one type of plantain but several,

from Guinea, Cape Verde, other parts of Africa, and even Andalusia (the Muslims cultivated bananas during their occupation of the peninsula, a practice that likely continued after their expulsion in 1492). Manuel de Paz Sánchez, "'Plantain of Guinea': The Atlantic Adventure of Banana," *Global Journal of Science Frontier Research* 14, no. 2 (2014): 49 and 60.

38. *Dela natural hystoria*: "que es cosa para no se creer sin verlo" (85 [chap. 80]).

39. *Dela natural hystoria*, 84 [chap. 80].

40. *Historia* (1535): "En la verdad ni este es árbol ni plátano saluo vna planta" (lxxix r [bk. 8, chap. 1]).

41. Pietro Martire d'Anghiera, *Décadas del nuevo mundo*, trans. Joaquin Torres Asensio (Madrid: Ediciones Polifemo, 1989): "aunque se diferencia muchísimo del plátano, y no tiene parentesco ninguno con él" (466 [dec. 7, chap. 9]).

42. *Historia* (1535), lxxx r [bk. 8, chap. 1].

43. Despite Oviedo's insistence that the fruit he describes is not a plantain, our contemporary nomenclature considers the plant as such.

44. *Historia* (1535), lxxx r [bk. 8, chap. 1].

45. *Historia* (1535), lxxx r [bk. 8, chap. 1].

46. *Historia*: "mensajero o significador" (ed. Tudela Bueso, 3:315 [bk. 29, chap. 26]).

47. Carrillo Castillo, *Naturaleza*, 304.

48. *Historia* (1535): "La qual por ser tan diferente y señalada hoja entre todas las otras la puse aquí" (lxxxiii r [bk. 8, chap. 13]).

49. *Historia* (1535): "Ay otra generación de yuca que las ramas y el fruto no es diferente de la que he dicho salvo en la hoja: porque aunque es de siete o nueve departiciones cada hoja, es de otra hechura" (lxxiii r [bk. 7, chap. 2]).

50. William E. Court, "The Ancient Doctrine of Signatures or Similitudes," *Pharmacy in History* 29, no. 3 (September 1999): 41.

51. *Historia* (1535), cxii r [bk. 15, prologue].

52. *Historia* (1535): "como si quisiessen enseñar a los hombres que son para curar las heridas de las tales lanzas o llagas" (cxvii v [bk. 11, chap. 5]).

53. *Historia* (1535), cxvii v [bk. 11, chap. 4].

54. *Historia* (1535): "y a mi parecer no está aquí mal contrahecha esta hoja sino muy bien, y por la misma mesura teniendo delante la misma y natural de esta excelente y salutífera planta" (cxvii v [bk. 11, chap. 4]).

55. *Historia* (1535): "La hoja de aquesta yerua es de la forma que aquí está dibuxada y de tal figura. Saluo que serán dos tantos mayores que está que va aquí pintada y algunas menores" (xcviii r [bk. 11, chap. 5]).

56. Martin Kemp, "Leonardo's Drawings for 'Il cavallo del Duca Francesco di bronzo': The Program of Research," in *Leonardo da Vinci's Sforza Monument Horse: The Art and the Engineering*, ed. Diane Cole Ahl (Bethlehem, PA: Lehigh University Press, 1995), 74.

57. *Historia* (1535), lxxvi v [bk. 7, chap. 8] and xci v–xcii r [bk. 10, chap. 1].

58. *Historia* (1535): "No la repito esta fruta porque tiene el mismo dibuxo que la guanábana de suso" (lxxxiiii r [bk. 8, chap. 18]).

59. *Historia* (1535): "Aquel matiz o sombra que está en las puntas destas hojas dibuxadas aueys de entender que es lo que tienen como morado" (xcviii r [bk. 11, chap. 5]).

60. *Historia* (1535), xcviii [bk. 11, chap. 5].

61. *Historia* (1535), lxxxiiii v [bk. 8, chap. 20].

62. In a passage in *Batallas* that Carrillo Castillo dates shortly before Oviedo's death in 1557, he mentions being in the process of writing a fourth part to the *Historia general*. This volume has never been located. Carrillo Castillo, introduction to *Oviedo on Columbus*, 28.

63. In Chapter 7, I provide a detailed history and analysis of the Monserrat manuscript's books on natural history.

64. Carrillo Castillo, *Naturaleza*, 109–111. Notably, the Truxillos manuscript entirely omits the natural history books of part 1. Several centuries later, Juan Bautista Muñoz observed that books 10, 13, 14, and 15 of part 1 of the Monserrat manuscript—on medicinal trees and plants, water animals, birds, and insects, respectively—had vanished.

65. In the letter that introduces the third volume of *Navigazioni e Viaggi* (1556), the Italian travel writer Giovanni Battista Ramusio describes the three-part *Historia general*—which he had never read, but about which he had communicated with Oviedo—as containing more than four hundred illustrations. This number is likely an exaggeration; although there are more illustrations in the Monserrat manuscript than in Oviedo's published writings on the Indies, extant portions of the manuscript do not suggest that it included as many as Ramusio alleges. Carrillo Castillo, *Naturaleza*, 244.

66. *Historia* (Monserrat): "Es hecha mas no como esta" (2:37r [bk. 7, chap. 2]).

67. In contrast, Carrillo Castillo interprets this statement to mean that Oviedo left such changes to the printer's discretion. *Naturaleza*, 317.

68. *Historia* (Monserrat): "Está por hazer a semejanza desta" (2:66v [bk. 11, chap. 5]).

69. *Historia* (1535), cxvii r [bk. 11, chap. 4].

70. Oviedo denominates as cinnamon a plant that has properties similar to those of the South Asian species. Turner conjectures that this pseudocinnamon might be the ishpingo, a South American tree. He also identifies the herb "y" as morning glory and the *perorica* possibly as wormroot. "Forgotten Treasure," 21 and 26.

71. *Historia* (Monserrat), 2:66v [bk. 9, chap. 26].

72. *Historia* (Monserrat), 2:68r [bk. 9, chap. 31].

Chapter 6

1. *Dela natural hystoria*: "porque a todos los hombres fuesen notorias tan grandes y maravillosas y nuevas cosas" (2).

2. Antonio de Villasante, "Unnamed," *Santo Domingo en los manuscritos de Juan Bautista Muñoz*, ed. Roberto Marte (Santo Domingo: Ediciones Fundación García Arévalo, 1981), 449. Villasante explains that in the province of Salvaleon, on Hispaniola, the natives refer to the plant as *boni*. In other provinces they use the name *guacuñasco* and on the mainland, *canaguey*.

3. Marcus Milwright, "The Balsam of Matariyya: An Exploration of a Medieval Panacea," *Bulletin of the School of Oriental and African Studies, University of London* 66, no. 2 (2003): 200–203.

4. Milwright, "Balsam," 193.

5. Milwright, "Balsam," 206.

6. In medieval Europe, balsam was used in the Christian practice of embalming and was an ingredient in chrism, the substance with which a monarch was anointed. Sultans often gifted balsam to rulers such as Charles VII of France (1403–1461), Lorenzo de' Medici (1449–1492), and the kings of Aragon. Milwright, "Balsam," 206–207.

7. "Asiento del bálsamo," April 4, 1528, AGI, Indiferente 421, L.13, F. 85r–86v.

8. "Carta del licenciado Barreda": "El engaño que Vuestra Majestad ha recibido en aprovar los dichos médicos por bálsamo un licor que el dicho Villasanta allá llevó" (October 26, 1528, AGI, Patronato 174, R. 43).

9. "Carta del licenciado Barreda," October 26, 1528, AGI, Patronato 174, R. 43.

10. "Carta del licenciado Barreda": "larga experiencia y mucho conocimiento" (October 26, 1528, AGI, Patronato 174, R. 43).

11. Of the nearly thirty royal charters issued in 1530 to encourage hospitals and medical professionals to experiment with the balsam and report their results, no such requests were sent to the New World. These numbers are a clear indication that the Crown did not seek input from overseas physicians, despite whatever potential knowledge they might have had of the tree. Barrera-Osorio, *Experiencing*, 21–22.

12. "Prohibición de coger y beneficiar el bálsamo," April 20, 1528, AGI, Indiferente 421, L. 13, F. 110r–111r.

13. "Real cédula al presidente y oídores de la Audiencia de la isla Española para que se informen sobre la conveniencia de mantener el asiento que se había concedido al vecino Antonio de Villasante," December 6, 1528, AGI, Indiferente 421, L. 14, F. 402v–403r.

14. "Real provisión a Antonio de Villasante," April 20, 1528, AGI, Indiferente 421, L.13, F. 250v–254v.

15. "Real cédula al obispo de Santo Domingo y la Concepción," April 5, 1530, AGI, Indiferente 422, L.14, F. 70r–71r; "Real cédula ordenando a todos los maestros, pilotos, marineros y pasajeros," April 5, 1530, AGI, Indiferente 422, L.14, F. 69r–70r.

16. "Real cédula ordenando a las justicias tengan cuidado de que los medicos no hagan declaraciónes," April 5, 1530, AGI, Indiferente 422, L.14, F. 67v–68r.

17. Barrera-Osorio, *Experiencing*, 21.

18. *Historia general* (Monserrat), 2:63r [bk. 11, chap. 4].

19. It is possible that Oviedo met the doctor in 1514; Barreda traveled as a passenger on Pedrarias Dávila's voyage to Castilla del Oro, and Oviedo as a member of the expedition. Barreda's final destination was Santo Domingo, where he ended up establishing a medical practice. Barrera-Osorio, *Experiencing*, 19.

20. In recounting his first encounter with the prickly pear, Oviedo refers to Barreda as a "gran medico" (*Historia* 1535, lxxxvi r [bk. 8, chap. 25]).

21. *Historia* (1535), xciii r [bk. 10, chap. 3].

22. *Historia* (1535): "Del balsamo verdadero Plinio lib.xii.capitu.xxvi. y otros auctores muchos han escripto, y no ay necessidad aqui de hablar en el pues los efetos del buen balsamo son apropriados a otras cosas muy apartadas delas que con este licor artificial se curan o quieren algunos curar" (xciiii v [bk. 10, chap. 3]).

23. *Historia* (1535), lxxxix r [bk. 9, chap. 10].

24. *Historia* (1535), lxxx r [bk. 8, chap. 1].

25. Álvarez López, "La historia natural," 579.

26. Ernst Schäfer, "Antonio de Villasante, descubridor droguista en la Isla Española," *Investigación y proceso* 9, no. 1 (1935): 13–15; Esteban Mira Caballos, "La medicina indígena en la Española y su comercialización (1492–1550)," *Asclepio* 49, no. 2 (1997): 185–198; Barrera-Osorio, *Experiencing*, 21.

27. *Historia* (1535), xciii v [bk. 10, chap. 3].

28. *Historia* (1535): "secreto" (xciii v [bk. 10, chap. 3]).

29. *Historia* (1535), xciii v [bk. 10, chap. 3].

30. *Historia* (1535): "Sea el inuentor de aqueste balsamo artificial qualquiera que aya seydo, qual que lo publico y gozo del interés primero fue este Antonio de Villasante: al qual la C.M. del emperador rey nuestro señor hizo mercedes por ello" (xciii v [bk. 10, chap. 3]).

31. *Historia* (1535): "Pero en esto yo me remito a los que lo han experimentado porque yo no lo he visto vsar ni exercer" (xciii v [bk. 10, chap. 3]).

32. *Historia* (1535): "He oydo grandes loores deste bálsamo y de sus operaciones. También he oydo a otros blassemarlo y dezir que es peligroso donde no se sabe aplicar" (xciii v [bk. 10, chap. 3]).

33. *Historia* (1535), xciii v [bk. 10, chap. 3].

34. *Historia* 1535, xciii v [bk. 10, chap. 3]. Oviedo describes the balsam as resembling a pear tree in Castile, while its leaf is shaped like that of a pomegranate, but thinner. Its bark is dry and burns well; its leaves are green and fresh. Like other tree resins, Oviedo reports that the bark has an aromatic scent when scratched.

35. *Historia* (1535): "y más pudiera dezir del segun me han informado y aun según lo que yo he visto de sus efectos a pro y a contra pero no quiero que nadie se cure por mis palabras, ni desseo tal credito en medicina, pues que no la estudie ni es de mi professión" (xciiii r–xciiii v [bk. 10, chap. 3]).

36. *Historia* (1535): "Pero desta agua ni del balsamo yo no me quiero entender a mas: pues hay aquí muchos que por esperiencia pueden hablar más largo en ello: y porque esta prohibido que ninguno lo haga" (xciiii r [bk. 10, chap. 3]).

37. *Historia* (1535): "y aqueste se murió sin cumplir lo que prometió. Pero yo digo lo que es publico: y cuanto al tesoro que auia de dar no se efetao. En verdad si mi parecer se tomasse ni su majestad pornía tal entredicho en cosa de que tanto bien podría resultar ni dexaría de mandar lo hazer a quantos quisiesen y después repartirlo por todos los que lo oviessen menester: pues que para el rey no pueden faltar otros interesses mayores para el acrecentamiento de sus rentas" (xciiii r [bk. 10, chap. 3]).

38. *Historia* (1535), xciiii r [bk. 10. chap. 3].

39. *Historia* (1535), cxvii r [bk. 11, chap. 4].

40. *Historia* (1535): "Creo que yo deve auer otras yerbas y plantas y árboles innumerables apropriados a nuestras passiones y llagas humanas" (cxvii v [bk. 11, chap. 5]).

41. *Historia* (1535): "que es mejor y más seguro que el bálsamo o licor de Villasante" (cxvii r [bk. 11, chap. 4]).

42. *Historia* (1535), cxvii r [bk. 11, chap. 4].

43. *Historia* (1535), xcviii r [bk. 11, chap. 5].

44. *Historia* (1535), cxvii r [bk. 11, chap. 4]. Oviedo describes the process of making the artificial balsam ointment as follows: "Toman los cogollos desta planta y aun algunos con ellos estos sus razimos y granos y hazen troços aquellos tallos y cuezenlos en agua hasta que mengua más de la mitad y más y más hasta que se espessa como vn arrope o miel, y después dexan lo assentar, y en llagas y desgarraduras aunque falte carne en la herida restaña la sangre y cura las llagas marauillosamente" (cxvii r [bk. 11, chap. 4]). He provides similar details with respect to the ointment made from the *perebecenuc*: "Cuzen vn puño de aquesta yerba digo de los tallos y hojas más tiernos tanta cantidad como se podrá apretar en vn puño o de grosseza de la muñeca de vn braço y después que de vn açunbre de agua que echen con aquel manojo de la yerba ha menguado la tercia parte apartan la olla en que se ha cozido desuiada del huego y dejan la hasta que esté quasi fría, y con aquella agua y vn paño limpio mojado en ella lavan la llaga" (*Historia* [1535], xcviii r [bk. 11, chap. 5]).

45. *Historia* (1535): "Según algunos afirman se haze de troços pequeños destos tales árboles que cozidos en agua sale dellos vn licor" (xciii v [bk. 10, chap. 3]).

46. *Historia* (1535), cxvii v [bk. 11, chap. 4]; xcviii r [bk. 11, chap. 5]. Oviedo describes the leaves of the artificial balsam as green, and "gruessas y anchas, y por de dentro son más verdes que por las espaldas: llamo yo las espaldas a la parte que tienen levantado o más releuado el neruio que va por mitad de la hoja desde el peçón a lo más alto o fin della. El qual peçón no es verde sino quasi colorado y las hojas en algunas partes dellas están alumbradas de una roxeza o color que tira a un roxo morado" (cxvii [bk. 11, chap. 4]). Alternatively, the *perebecenuc*, "tiene muchas hojas anchas y agudas en las puntas y en el talle quieren parecer hierros de lanças ginetas de los pequeños como si quisiesen enseñar a los hombres que son para curar las heridas de las tales lanças o llagas. Son muy delgadas en si estas hojas y verdes y las puntas dellas algo moradas y los astiles o tallos en que nacen estas hojas son assi mismo algo morados y de la color de las puntas de las hojas, aunque algunas ay que no son puntiagudas y son algo más romas, pero las vnas y las otras tienen los estremos de aquella color entre leonado y morado" (cxvii v [bk. 11, chap. 5]).

47. *Historia* (1535): "y a mi parecer no está aquí mal contrahecha esta hoja sino muy bien" (cxvii v [bk. 11, chap. 4]).

48. The primary cause of these civil wars (1520–1521) was the tensions that developed between Charles V's Burgundian administration and Castilian cities. At the Cortes in 1517 and 1520, the monarch had agreed to appoint Spanish natives to government positions, but ultimately he did not adopt these measures. Furthermore, he modified the Castilian tax code to require the previously tax-exempt hidalgo class to pay a municipal subsidy and certain churches to pay an additional tax. As a final afront, in May 1520, the king left Spain for the Holy Roman Empire and appointed Adrian of Utrecht as regent of Spain. Aurelio Espinosa, *The Empire of the Cities: Emperor Charles V, the Comunero Revolt, and the Transformation of the Spanish System* (Boston: Brill, 2009), 65–84.

49. *Batallas*, ed. Avalle-Arce, 214 [bat. 2, quin. 2, diál. 1].

50. Pliny, *Naturalis historia*, bk. 21, chap. 57.

51. Sebastián de Covarrubias Orozco, "cardo," in *Tesoro de la lengua castellana o española* (Madrid: Melchor Sánchez, 1674), https://www.cervantesvirtual.com/, accessed October 5, 2022.

52. *Historia* (1535): "vna manera de cardos" (lxxxv v [bk. 8, chap. 24]).

53. Del Weniger, *Cacti of the Southwest: Texas, New Mexico, Oklahoma, Arkansas, and Louisiana* (Austin: University of Texas Press, 1969), 4.

54. *Historia* (1535): "No tienen hoja saluo vnas ramas o braços luengos" (lxxxv v [bk. 8, chap. 23]).

55. *Historia* (1535): "Estos mismos cardos se convierten en aquellos árboles" (lxxxvi r [bk. 8, chap. 25]).

56. *Historia* (1535): "Porque en algunas cosas pintadas en papel yo la tengo puesta más ha de seys años" (xcii r [bk. 10, chap. 1]).

57. Michael DeFelice, "Prickly Pear Cactus, *Opuntia* spp.: A Spine-Tingling Tale," *Weed Technology* 18, no. 3 (2004): 870; Weniger, *Cacti*, 60.

58. DeFelice, "Prickly Pear," 869.

59. *Historia* (1535): "A mí no me duele nada mas daría yo mi caballo y otros quatro por estar cerca de Santo Domingo o del licenciado Barreda que es gran medico, porque sin dubda yo devo de tener rotas quantas venas tengo en el cuerpo" (lxxxvi v [bk. 8, chap. 25]).

60. *Historia* (1535): "Es tal burla" (lxxxvi v [bk. 8, chap. 25]).

61. *Historia* (1535), xcii r [bk. 10, chap. 1].

62. *Historia* (1535): "Ni yo le sé el nombre que los indios le dan en esta ysla ni en las otras, ni en la tierra firme, donde en cada parte se nombra en diferenciada manera por la gran diferencia e moltitud delas lenguas que en estas Indias ay" (xci v [bk. 10, prologue]).

63. Like the welding tree, Oviedo observes that the pineapple is widespread in the Indies and goes by many different names. He refers to the plant as *piña* because it is known as such among his compatriots, although he considers the name to be a misnomer. Oviedo does not claim ignorance of the different indigenous names used for the plant, as he does in the case of the welding tree, but rather describes several popular varieties. Rabasa, *De la invención*, 167.

64. *Historia* (1535): "Las quales ramas fueron primero hojas e pencas cada vna dellas, e de aquella hoja o penca nascieron otras, e de las otras, otras" (xci v [bk. 10, chap. 1]).

65. *Historia* (1535): "mostruo del genero delos arboles" (xcii r [bk. 12, chap. 1]).

66. *Historia* (1535): "Es mas para verle pintado de mano de Berruguete o otro excelente pintor como el o aquel Leonardo da Vince o Andrea Mantheña famosos pintores que yo conocí en Ytalia: que no para darle a entender con palabras" (xci v [bk. 10, prologue]).

67. *Historia* (1535): "Pero con mi mal debuxo porné aquí la forma que tiene este árbol si lo supiere hazer: para que juntamente con lo que del tengo dicho mejor se pueda entender y considerar" (xcii r [bk. 10, chap. 1]).

68. *Historia* (1535), xcii r–xcii v [bk. 10, chap. 1].

69. William F. Keegan, "The 'Classic' Taíno," in *The Oxford Handbook of Caribbean Archaeology*, ed. William F. Keegan, Corinne L. Hofman, and Reniel Rodríguez Ramos (Oxford: Oxford University Press, 2013), 72.

70. *Historia* (1535), cii v [bk. 13, chap. 3].

71. *Historia* (1535), xvii v [bk. 2, chap. 12] and cxv r [bk. 15, chap. 4]. Oviedo identifies a hog plum that he claims Martire incorrectly calls a *mirobálano*. He also conjectures that the island of Fernandina must be the same one that Martire refers to as Alpha/o or Juana. So too, he disagrees with the Italian chronicler's account of a tree so bitter that shipworms do not eat its wood, arguing that he knows of no wood with that property.

72. *Historia* (1535): "Pedro Mártir . . . dize que estas yvanas son semejantes al cocodrilo, que son animales del río nilo" (ciii r [bk. 13, chap. 3]).

73. *Historia* (1535): "y si aqui me he alargado tanto ha sido para sacar a los letores de la opinion de Pedro Mártir" (ciii r [bk. 13, chap. 3]).

74. Anglería, *Décadas*, 33 [dec. 1, chap. 3]. "Las cuales afirman que en nada absolutamente se diferencian de los cocodrilos de Egipto sino en el tamaño: pues de los cocodrilos dice Plinio que se encontraron algunos de dieciocho codos, pero las mayores de estas serpientes tienen ocho pies."

75. *Historia* (1535), cii v [bk. 13, chap. 3].

76. Isidore of Seville, *Etymologies*, 260 [bk. 12, chap. 6, sec. 19]; Pliny, *Naturalis historia*, bk. 8, chap. 37.

77. *Historia* (1535): "Él se engañó mucho y a semejantes y notorios errores están obligados los que estas cosas escriben por oydas, porque estas yvanas no son mayores animales de lo que tengo dicho" (ciii r [bk. 13, chap. 3]).

78. Although Oviedo denies any similarities between the crocodile and the iguana, it is possible that part of his confusion as to how to classify the New World species stems from the lack of consensus among classical and medieval authorities as to the crocodile's identity.

Pliny the Elder classifies the creature as a terrestrial animal, whereas Isidore of Seville considers it a fish.

79. *Dela natural hystoria*: "De las bíuoras y culebras y sierpes y lagartos y sapos y otras cosas semejantes" (61). Although Oviedo mentions the iguana in chapter 6 of the *Sumario*, it remains nameless and uncategorized until chapter 56. Jeremy Paden argues that this omission signals the difficulty of making the Indies intelligible for Charles V. Paden, "Iguana," 205.

80. *Historia* (1535): "No se saben determinar los Christianos si son de genero de carne o de pescado" (cii v [bk. 12, chap. 3]).

81. Pliny, *Naturalis historia*, bk. 8, chap. 49 and bk. 8, chap. 48.

82. Isidore of Seville, *Etymologies*, 259 [bk. 12, chap. 6, sec. 3].

83. Pietro Martire recounts the initial reaction of Columbus and his men to the appearance and taste of the iguana in chapter 5 of his first *Década* (1516). He writes, "Aprendieron más tarde que esas serpientes nacen en esta isla, y los nuestros hasta ahora no se habían atrevido a gustarlas por su fealdad, que parecía causar horror, no sólo asco. El Adelantado, inducido por el gracejo de la hermana del cacique, determine catarlas poco a poco; pero apenas el saber de aquella carne comenzó a gustar al paladar y garganta parecía que las deseaba a boca llena. Después ya no las probaba con la punta de los dientes o aplicando apenas los labios, sino que, habiéndose hecho todos glotones, de nada hablaban ya sino del grato sabor de las serpientes y de que tales viandas eran más exquisitas que entre nosotros las de pavo, faisán y perdiz" (52 [dec. 1, chap. 5]).

84. Paden attributes these revisions not to the Monserrat manuscript but rather to the second edition of the first part of the *Historia*, published in 1547 (not 1548, as Paden alleges). "Iguana," 205. However, Daymond Turner's side-by-side comparison of the two print editions found only slight differences between them, mainly orthographic. "Aborted," 110. In the *Historia* (1547), the chapter on the iguana is still located in *Book 13*, and Oviedo continues to express the same doubts regarding its identity.

85. *Historia general* (Monserrat): "Yo le avría por carne. Lo qual no digo para que ninguno dexe de seguir su voluntad, y principalmente la del perlado y lo que la Iglesia ordenare" (310v [bk. 12, chap. 7]).

86. *Historia* (1535): "Esta que aquí yo pinte quiere algo semejar a la yvana y a la verdad esta forma tienen" ciii v (bk. 13, chap. 3]).

87. *Historia* (1535): "con la protestación o auiso que al principio deste capítulo se toco, que aqueste animal es como neutral y que no se determinan los hombres si es terrestre o de agua" (ciii v [bk. 13, chap. 3]).

Chapter 7

1. *Dela natural hystoria*, 103. Oviedo claims in the concluding lines of the *Sumario* that he had almost finished a preliminary version of the *Historia general*. While this statement should not be taken too seriously, at the very least it indicates that the outlines of a subsequent project were beginning to take form.

2. Carrillo Castillo, *Naturaleza*, 130. It is possible that Oviedo continued to write yet a fourth part to the *Historia general* upon his return to Santo Domingo, although its manuscript has never been located.

3. In *Hijos ilustres de Madrid* (1790), José Antonio Álvarez y Baena communicates news of the *Historia general*'s impending publication, which he credits to the efforts of don Francisco Cerdá y Rico, librarian of the Biblioteca Real, who began to collect copies and originals of

all fifty books in 1775. Remedios Contreras, "Intentos de publicación de la 'Historia general y natural de las Indias' de Fernández de Oviedo anteriores a Amador de los Ríos," in *América y la España del Siglo XVI*, ed. Francisco de Solano and Fermín del Pino (Madrid: Instituto Gonzalo Fernández de Oviedo, 1982), 1:126. Jesús Carrillo Castillo, "The *Historia general y natural de las Indias* by Gonzalo Fernández de Oviedo," *Huntington Library Quarterly* 65, nos. 3/4 (2002): 321.

4. Criticism of the modern editions of the *Historia general* is not new; José Anadón points to several general oversights in Ríos's editorial work with the manuscript. Karl Kohut calls for a critical edition of the work that takes into consideration the development of the manuscript over time. José Anadón, "Los manuscritos originales de la *Historia general y natural* de Fernández de Oviedo," *Revista de la Universidad de México* (August 1979): 31. Karl Kohut, "Fernández de Oviedo, historiador y literato: Humanismo, christianismo, e hidalguía," in *Historia y ficción: Crónicas de América*, coord. Ysla Campbell (Ciudad Juárez, MX: Universidad Autónoma de Ciudad Juárez, 1992), 49n17.

5. Contreras, "Intentos," 120–121.

6. Most importantly, Ríos does not often differentiate between the base text of the Monserrat manuscript and later marginal additions and edits that Oviedo made to it. Anadón, "Los manuscritos originales," 31. Other decisions that Ríos made also impacted reception of the work: he did not include the notes Oviedo left for the printer regarding the work's hand-drawn illustrations, which he removed from alongside the text and placed in an appendix.

7. The Monserrat manuscript has suffered many losses during the nearly five centuries of its existence. This history can be summarized as follows: when Juan Bautista Muñoz consulted the manuscript for the first time in 1780, he recorded absences in all four volumes. Still more disappearances occurred at the beginning of the twentieth century. Several secondary sources record some of the material in these missing portions of the manuscript. In the 1560s, the Spanish inquisitor Andrés Gascó copied certain sections of the Monserrat manuscript (housed in the Palacio Real, the Biblioteca de la Real Academia de Historia in Madrid, and the Biblioteca Colombina in Seville); however, he did not transcribe most of the first part of the *Historia general*, including its natural history. In the eighteenth century, Juan Bautista Muñoz copied other parts of the Monserrat manuscript whose originals have since been lost. I provide more details shortly concerning the losses experienced by the Monserrat manuscript's books on natural history. Carrillo Castillo, *Naturaleza*, 109–110.

8. Carrillo Castillo, *Naturaleza*.

9. Carrillo Castillo, *Naturaleza*, 114.

10. Carrillo Castillo, *Naturaleza*, 119. Carrillo Castillo suggests that this decision could be related to Oviedo's failure to publish a revised part 1 and part 2. He maintains that it is only at this point that Oviedo turned his sights to the completion and publication of the three-part *Historia general*.

11. Carrillo Castillo, *Naturaleza*, 109.

12. Carrillo Castillo, *Naturaleza*, 112.

13. Although further losses took place prior to Ríos's work with the manuscript, these did not affect any of the remaining natural history books.

14. *Book 17* and *Book 19* are incomplete. The manuscript of the former begins partway through chapter 9 and ends in chapter 28; that of the latter contains the first four chapters of the book as well as part of the fifth.

15. The only book in part 1 to include more new additions is *Book 6*, to which Oviedo added forty new chapters, most of them likely in the final stages of composing the manuscript (1542–1549).

16. As Oviedo informs the viceroy don Antonio de Mendoza in a letter dated March 1, 1542—which he transcribed in the *Historia general*—by this date he had already written fifty books, divided into three parts or volumes. He mentions too that he had expanded the first part of the work by almost a third. Ed. Tudela Bueso, 4:254 [bk. 33, chap. 53]. A study of the extant portions of the Monserrat manuscript and its table of contents suggests that the books on natural history (6–15) experienced most of this growth. Myers, "Imitation," 524n3. By 1550, when Oviedo described the manuscript to Bishop Pedro la Gasca, it was no longer divided into three volumes but four. Carrillo Castillo, *Naturaleza*, 116. These later expansions to the manuscript are related to the ongoing Spanish conquest and colonization of the Americas. Although Oviedo's informants occasionally provide him with information on natural phenomena, for the most part these observations are secondary to other concerns.

17. Oviedo conceived of *Book 6: Of Deposits* as dedicated to phenomena so unique as to be deserving of special mention. Entries include reports on flora and fauna interspersed with accounts of indigenous cultural practices, weather-related phenomena, and other topics. Oviedo revisits many of these subjects later, in the sections of the *Historia general* corresponding to the geographies in which they were observed or took place. The abundant cross-referencing in *Book 6* suggests that many of its chapters were later additions to the manuscript. Given this chronology, as well as the eclectic nature of the topics it addresses, I do not consider *Of Deposits* as representative of the general trends that mark natural history at large in the Monserrat manuscript. Vanina Teglia analyzes additional features of *Book 6* that set it apart from others on New World nature: "Claroscuros del archive colonial: La escritura sobre la naturaleza de Fernández de Oviedo," *Huarte de San Juan: Geografía e Historia* 27 (2020): 267–290.

18. Carrillo Castillo, *Naturaleza*, 117.

19. Carrillo Castillo, *Naturaleza*, 120.

20. Oviedo, *Historia general*: "Agora que es llegado el tiempo en que esta primera parte enmendada e acrescentada se reimprime (e también la segunda e tercera) hame parecido que porque las materias anden juntas, que se pongan de manera que el letor no ande a buscar mis promesas" (ed. Tudela Bueso, 1:260 [bk. 8, chap. 20]).

21. *Historia general*, ed. Tudela Bueso, 1:260 [bk. 8, chap. 20].

22. *Historia* (1535): "Ay mucho mas que notar en todas estas cosas y en cada vna de ellas y en otras que para allí se guardan porque competen a aquella tierra y no a esta" (lxxxvii v [bk. 9, prologue]).

23. *Historia* (1535): "En la segunda parte de esta natural y general historia de indias (y enla tercera) se hallaran otras muchas particularidades y nueuas plantas y yerbas que reseruo para en sus proprios lugares y provincias" (xcv v [bk. 11, prologue]).

24. *Historia* (1535): "En la segunda y tercera partes quando se trate de las cosas de tierra firme aura mucho más que escriuir desta materia porque alla ay muchos y muy diferentes animales de todos los de nuestra España" (xcviii v [bk. 12, prologue]).

25. *Historia general* (Monserrat): "Mas por no dilitar y por que la experiencia sea moderada y no se atienda para saber lo que en este tiempo está sabido en estas indias me paresce que será bien que lo que tocare a estos árboles salvajes se ponga aquí en este libro . . . porque la materia esté junta y no desembrada ni la segunda ni la tercera parte destos libros la dividan" (2:41r [bk. 9, prologue]).

26. *Historia general* (Monserrat): "Se dirá de algunas plantas y yervas medicinales y provechosas que la natura produze en esta y otras islas y en la tierra firme" (2:60r [bk. 11, prologue]).

27. *Historia general* (Monserrat): "Fecho esto tractare de todos los otros animales (en quien no se habló en la primera impresión que tuvo esta primera parte dela general historia destas indias) . . . en cuales quier ylas o partes dela tierra firme dando a cada animal su propia patria . . . porque el jaez de la materia y calidad de que aqui se tratare se halle junto" (306r [bk. 12, prologue]).

28. *Historia general* (Monserrat): "Pienso yo que es la misma corbana de la qual madera negra y de su perpetuydad debaxo de tierra se dixo en el libro precedente ~~adelante~~ cuando se trata de los árboles de cacao que también se llaman caraguat" (2:62r [bk. 9, chap. 16]).

29. Oviedo edited several key passages in *Book 7*'s prologue in the Monserrat manuscript sometime after its initial composition. Given that some of the text has been obscured or deleted, it is impossible to know the nature of those changes.

30. Carrillo Castillo, *Naturaleza*, 120.

31. Oviedo's account of corn (bk. 7, chap. 1) in the *Historia* (1535) already acknowledges that making cornbread is a common practice for many indigenous communities on the islands and mainland, so the passage remains unchanged in the Monserrat manuscript. Similarly, Oviedo determines that the use of the canoa is widespread in the Indies; in the *Historia* (1535), and then later in the Monserrat manuscript, he refers to this as a form of transportation used "on this island Hispaniola and in all the other parts of these Indies that until the present moment are known" (lxi r [bk. 6, chap. 4]).

32. To provide just one example, Oviedo alters the chapter title "Of the robles on the island of Hispaniola" from the *Historia* (1535) (bk. 9, chap. 8 [lxxxviii v]) to read, "Of the roble on the island of Hispaniola and in other parts of the Indies" in the Monserrat manuscript (bk. 9, chap. 8 [2:46v]).

33. *Historia general* (Monserrat), 2:48r–2:50r [bk. 9, chap. 11]; 308b–309a [bk. 12, chap. 5].

34. *Historia general* (Monserrat), 2:41v–2:45v [bk. 9, chap. 4]. Beyond infrequent updates to chapter titles, there are no substantive revisions to *Book 6* and *Book 7* in the Monserrat manuscript. In many instances, this is likely because in the *Historia* (1535) these books already referenced a wider geographical sphere—many of the chapter headings cite not just Hispaniola but also the neighboring islands and *tierra firme*. For this reason, there was no need for Oviedo to amend these chapters as he did others that had been focused exclusively on Hispaniola in the *Historia* (1535).

35. As quoted in Carrillo Castillo's *Naturaleza e imperio*: "discurriese por aquellas tierras donde no ha andado para ver lo que no tiene visto" (76).

36. As quoted in Carrillo Castillo's *Naturaleza e imperio*: "cumplidamente y por buen estilo" (77).

37. Carrillo Castillo, *Naturaleza*, 76–77. Carrillo Castillo argues that Oviedo, at age fifty-four in 1532, was no longer willing to travel extensively, preferring to carry out his task from the comforts of his home in Santo Domingo.

38. Carrillo Castillo, *Naturaleza*, 79. Three months after his appointment was confirmed, Oviedo requested that the Council of the Indies provide him with access to archival records and require captains and governors (*adelantados*) to give him accounts of their expeditions.

39. Carrillo Castillo, *Naturaleza*, 129. Carrillo Castillo observes that the most significant changes to the Monserrat manuscript between 1542 and 1549 concern the continued advances in

the colonial enterprise, news of which Oviedo obtained in many cases from witnesses to and participants in these events. The nature of these additions responds directly to the kind of testimony that Oviedo was able to acquire about the regions in the Americas where he had never been, both in the 1540s and earlier, in the 1530s.

40. Scholars such as Alberto M. Salas and Louise Bénat-Tachot have highlighted some of the key differences between part 1 of the *Historia general* and parts 2 and 3, although they have not queried their implications for the composition of the Monserrat manuscript. Salas disputes Edmund O'Gorman's characterization of parts 2 and 3 of the *Historia general* as a distancing of Oviedo from natural history, as he became more invested in the events of the conquest and colonization; he argues that Oviedo's lack of direct knowledge of these lands prevented him from engaging with New World nature in the same way as he had in his published works. Similarly, Bénat-Tachot argues that experience loses its place of privilege in the second and third parts of the work, given that Oviedo had to rely on other witnesses to supply him with information. Although Bénat-Tachot focuses her analysis primarily on the Amerindian in Oviedo's narrative, nevertheless her observations apply as well to the flora and fauna of the Indies. Salas, *Tres cronistas*, 59 and 107; Louise Bénat-Tachot, "La experiencia," 250.

41. New Castile was a swath of territory reaching from the Pacific Ocean to the Atlantic, which includes most of present-day Peru and Ecuador and parts of Brazil, Columbia, and Bolivia.

42. *Historia general* (Monserrat): "En aquella tierra he sabido de los capitanes Juan de Junco y Gomez de Corral y de otras personas que se deven creer que ay muchos osos de los mismos que hay en nuestra España" (327v [bk. 12, chap. 37]); *Historia general* (Monserrat): "Dizen los que por aquella tierra anduvieron que ay un animal" (328r [bk. 12, chap. 38]); *Historia general* (Monserrat): "según me han certificado los que en aquella tierra del norte han andado" (328v [bk. 12, chap. 40]).

43. *Historia general* (Monserrat), 326r [bk. 12, chap. 30].

44. These species include those that Oviedo identifies as the *maguey* or agave (chapter 11), *ymocona* (chapter 15), *guayero* (chapter 16), *cavallo* (chapter 17), *fesoles* or beans (chapter 18), and *names* (chapter 19).

45. Oviedo does not mention how he became acquainted with two other new species of trees—what he refers to as the *agorero* on the coast of the mainland close to the island of Cubagua and a particular kind of resin-producing tree in New Castile (chapters 29 and 30, respectively). Since Oviedo usually discloses the sources of this kind of second-hand information, it is likely that he had experienced these plants, or products derived from them, at some point in his life.

46. Castilla del Oro constituted the Caribbean coasts of southern Central America and northwestern South America.

47. *Dela natural hystoria*, 13 [vii].

48. Oviedo also describes many of these species in parts 2 and 3 of the *Historia general* as components of the ongoing narrative of conquest and colonization. He often copies the information included in these later sections from the corresponding entries in part 1, as he does in the case of the *bivana* (bk. 24, chap. 13) and the buffalo (*vacas montesas*) (bk. 36, chap. 2).

BIBLIOGRAPHY

Unpublished Documentary Sources

Archivo General de Indias, Seville, Spain

Indiferente, 421, L.13, f85r–86v.

April 4, 1528. "Asiento del bálsamo."

Indiferente, 421, L.13, f110r–111r.

April 20, 1528. "Real provisión a Antonio de Villasante."

Indiferente, 421, L.13, f110r–111r.

April 20, 1528. "Prohibición de coger y beneficiar el bálsamo."

Indiferente, 421, L.14, f402v–403r.

December 6, 1528. "Real cédula al president y oídores de la Audiencia de la isla Española para que se informen sobre la conveniencia de mantener el asiento que se había concedido al vecino Antonio de Villasante."

Indiferente, 422, L.14, f69r–70r.

April 5, 1530. "Real cédula ordenando a todos los maestros, pilotos, marineros y pasajeros . . ."

Indiferente, 422, L.14, f70r–71r.

April 5, 1530. "Real cédula al obispo de Santo Domingo y la Concepción."

Indiferente, 422, L.14, f67v–68r.

April 5, 1530. "Real cédula ordenando a las justicias tengan cuidado de que los medicos no hagan declaraciónes."

Patronato 174, f43.

October 26, 1528. "Carta del licenciado Barreda."

Fernández de Oviedo, Gonzalo. *Adición al libro 14 de la segunda parte* [Monserrat manuscript]. ca. 1535–1549. Biblioteca de la Real Academia de la Historia (RAH), Madrid, Spain. 9/6480 (14).

———. *Historia general de las Indias* [Monserrat manuscript]. ca. 1535–1549. The Huntington Library, San Marino, CA. 2 vols [books 6, 7, 9, 11]. Huntington Library. MSS HM 177. https://catalog.huntington.org/record=b1800280. Accessed August 19, 2021.

———. *Historia general de las Indias* [Reese Manuscript]. ca. 1535–1549. Book 12. Anonymous owner.

———. *Tercera parte de la Historia natural y general de las Indias* [Truxillos manuscript]. Biblioteca del Palacio Real, Madrid, Spain. 2/3042.

Published Sources

Adorno, Rolena. *Colonial Latin American Literature: A Very Short Introduction*. Oxford: Oxford University Press, 2011.

Adorno, Rolena, and Patrick Charles Pautz. *Álvar Núñez Cabeza de Vaca: His Account, His Life, and the Expedition of Pánfilo de Narváez*. 3 vols. Lincoln: University of Nebraska Press, 1999.

Albala, Ken. *Eating Right in the Renaissance*. Berkeley: University of California Press, 2002.

Alonso de Herrera, Gabriel. *Obra de agricultura*. Edited by José Urbano Martínez Carrera. Madrid: Ediciones Atlas, [1513] 1970.

Alvarez, Mari-Tere. "Artistic Enterprise and Spanish Patronage: The Art Market During the Reign of Isabel of Castile (1474–1504)." In *Art Markets in Europe, 1400–1800*, edited by Michael North and David Ormrod, 45–59. Aldershot: Ashgate, 1998.

Álvarez López, Enrique. "La historia natural en Fernández de Oviedo." *Revista de Indias* 17, no. 69 (1957): 541–601.

———. Introduction and notes to *De la natural historia de las Indias*, by Gonzalo Fernández de Oviedo, 5–40 and 197–223. Madrid: Editorial Summar, [1526] 1942.

———. "Plinio y Fernández de Oviedo." *Anales de ciencias naturales* (1940): 40–61.

Álvarez Peláez, Raquel. *La conquista de la naturaleza americana*. Madrid: Consejo Superior de Investigaciones Científicas, 1993.

Ambrosoli, Mauro. *The Wild and the Sown: Botany and Agriculture in Western Europe, 1350–1850*. Cambridge: Cambridge University Press, 1997.

Ames-Lewis, Francis. *Isabella and Leonardo: The Artistic Relationship Between Isabella d'Este and Leonardo da Vinci, 1500–1506*. New Haven, CT: Yale University Press, 2012.

Anadón, José. "Los manuscritos originales de la *Historia general y natural* de Fernández de Oviedo." *Revista de la Universidad de México*, August 1979, 30–31.

Anghiera, Pietro Martire d'. *Décadas del nuevo mundo*. Translated by Joaquin Torres Asensio. Madrid: Ediciones Polifemo, 1989.

Arias Martínez, Manuel. "Master of the Retablo." In *Alonso Berruguete: First Sculptor of Renaissance Spain*, edited by C. D. Dickerson III and Mark McDonald, 86–115. Washington, DC: National Gallery of Art, 2019.

Augustine. *Confessions*. Edited and translated by Carolyn J. B. Hammond. Cambridge, MA: Harvard University Press, 2016.

Avalle-Arce, Juan Bautista. "Oviedo a media luz." *Nueva Revista de Filología Hispánica* 29, no. 1 (1980): 138–151.

Ballesteros Gaibrois, Manuel. *Gonzalo Fernández de Oviedo*. Madrid: Fundación Universitaria Española, 1981.

Baraibar, Álvaro. "Estudio preliminar." In *Sumario de la natural historia de las Indias*, by Gonzalo Fernández de Oviedo, 9–59. Madrid: Iberoamericana Vervuert, [1526] 2010.

Barrera-Osorio, Antonio. *Experiencing Nature: The Spanish American Empire and the Early Scientific Revolution*. Austin: University of Texas Press, 2006.

Bataillon, Marcel, and Edmundo O'Gorman. *Dos concepciones de la tarea histórica con motivo de la idea del descubrimiento de América*. Mexico City: Imprenta Universitaria, 1955.

Bauman, Johanna Elizabeth. "Piero de' Crescenzi's *Liber ruralium commodorum*: Unearthing the Origins of the Pleasure Garden." PhD diss., University of Virginia, 2000.

Baxandall, Michael. *Giotto and the Orators: Humanist Observers of Painting in Italy and the Discovery of Pictorial Composition, 1350–1450*. Oxford: Clarendon Press, 1971.

Bénat-Tachot, Louise. "La experiencia en el proceso cognitivo de las sociedades indígenas en la *Historia general y natural de las Indias* de Fernández de Oviedo." In *Pensamiento eu-*

ropeo y cultura colonial, edited by Karl Kohut and Sonia V. Rose, 247–261. Madrid: Iberoamericana, 1997.

Blair, Ann. "Natural Philosophy." In *The Cambridge History of Science*, edited by Katharine Park and Lorraine Daston, 3.363–406. Cambridge: University of Cambridge Press, 2006.

Bleichmar, Daniela. *Visible Empire: Botanical Expeditions and Visual Culture in the Hispanic Enlightenment*. Chicago: University of Chicago Press, 2012.

———. *Visual Voyages: Images of Latin American Nature from Columbus to Darwin*. New Haven, CT: Yale University Press, 2017.

Blunt, Wilfrid, and Sandra Raphael. *The Illustrated Herbal*. New York: Thames and Hudson, [1979] 1994.

Bolaños, Álvaro Félix. "The Historian and the Hesperides: Fernández de Oviedo and the Limitations of Imitation." *Bulletin of Hispanic Studies* 72 (1995): 273–288.

Brendecke, Arndt. *The Empirical Empire: Spanish Colonial Rule and the Politics of Knowledge*. Berlin: De Gruyter, 2016.

Brown, Clifford M., and Sally Hickson. "Caradosso Foppa (ca. 1452–1526/27)." *Arte Lombarda* 119, no. 1 (1997): 9–39.

Brown, Jonathan. *Painting in Spain, 1500–1700*. New Haven, CT: Yale University Press, 1998.

Brunfels, Otto. *Herbarum vivae eicones*. Strasbourg: Johann Schott, 1530. https://www.metmuseum.org/art/collection/search/338754. Accessed November 2, 2022.

Bylebyl, Jerome J. "The School of Padua: Humanistic Medicine in the Sixteenth Century." In *Health, Medicine and Morality in the Sixteenth Century*, edited by Charles Webster, 335–370. Cambridge: Cambridge University Press, 1979.

Calkins, Robert G. "Piero de' Crescenzi and the Medieval Garden." In *Medieval Gardens*, edited by Elizabeth Blair MacDougall, 155–174. Washington, DC: Dumbarton Oaks, 1986.

Cañizares-Esguerra, Jorge. *Nature, Empire, and Nation: Explorations of the History of Science in the Iberian World*. Redwood City, CA: Stanford University Press, 2006.

Carrillo Castillo, Jesús. "The Eyes of the New Pliny: The Use of Images in Gonzalo Fernández de Oviedo's *Historia general y natural de las Indias*." In *The Art of Natural History: Illustrated Treatises and Botanical Paintings, 1400–1850*, edited by Therese O'Malley and Amy R. W. Meyers, 109–125. Washington, DC: National Gallery of Art, 2008.

———. "From Mt Ventoux to Mt Masaya: The Rise and Fall of Subjectivity in Early Modern Travel Narrative." In *Voyages and Visions: Toward a Cultural History of Travel*, edited by Jás Elsner and Joan-Paul Rubiés, 57–73. London: Reaktion Books, 1999.

———. "The *Historia general y natural de las Indias* by Gonzalo Fernández de Oviedo." *Huntington Library Quarterly* 65, nos. 3/4 (2002): 321–344.

———. Introduction to *Oviedo on Columbus*, 9–35. Edited by Jesús Carrillo Castillo. Translated by Diane Avalle-Arce. Turnhout, Belgium: Brepols, 2000.

———. "Naming Difference: The Politics of Naming in Fernández de Oviedo's *Historia general y natural de las Indias*." *Science in Context* 16, no. 4 (December 2003): 489–504.

———. *Naturaleza e imperio: La representación del mundo natural en la "Historia general y natural de las Indias" de Gonzalo Fernández de Oviedo*. Madrid: Fundación Carolina, 2004.

———. "'The World Is Only One and Not Many': Representation of the Natural World in Imperial Spain." In *Spain in the Age of Exploration, 1492–1819*, edited by Chiyo Ishikawa, 139–157. Lincoln: University of Nebraska Press, 2004.

Caruso, Carl. *Adonis: The Myth of the Dying God in the Italian Renaissance*. London: Bloomsbury, 2013.

Casas, Bartolomé de las. *Historia de las Indias.* Edited by Juan Pérez de Tudela Bueso. 2 vols. Madrid: Edición Atlas, 1961.

———. *Opúsculos, cartas y memoriales.* Edited by Juan Pérez de Tudela Bueso. Vol. 5. Madrid: Edición Atlas, 1958.

Cermenati, Mario. "Un diplomatico naturalista del rinascimento, Andrea Navagero." *Nuovo Archivio Veneto* 25, no. 1 (1912): 164–215.

Chambers, David, Jane Martineau, and Rodolfo Signorini. "Mantegna and the Men of Letters." In *Andrea Mantegna*, edited by Jane Martineau, 8–30. New York: Metropolitan Museum of Art, 1992.

Christiansen, Keith. *The Genius of Andrea Mantegna.* New York: Metropolitan Museum of Art, 2009.

Cicero, Marcus Tullius. *De natura deorum.* Translated by Francis Brooks. London: Methuen, 1896. https://oll.libertyfund.org/titles/539. Accessed April 9, 2020.

Clark, Kenneth. Introduction to *Leonardo da Vinci: Nature Studies from the Royal Library at Windsor Castle*, by Carlo Pedretti, 9–12. New York: Johnson Reprint, 1981.

Coello de la Rosa, Alexandre. "El proceso de escritura del *Sumario* (1526) de Gonzalo Fernández de Oviedo y Valdés en el seno de circulaciones y transferencias culturales con el humanismo italiano." *Pedralbes* 36 (2016): 143–178.

Coffin, David. "The Gardens of Venice." In *Magnificent Buildings, Splendid Gardens*, edited by Vanessa Bezemer Sellers, 196–199. Princeton, NJ: Princeton University Press, 2008.

Colección de documentos inéditos relativos al descubrimiento, conquista y organización de los antiguas posesiones españolas de América y Oceanía. Madrid: M. B. de Quirós, 1864–1884.

Colmeiro, Manuel. *Cortes de los antiguos reinos de León y de Castilla.* Madrid: Impresores de la Real Casa, 1884.

Contreras, Remedios. "Intentos de publicación de la 'Historia general y natural de las Indias' de Fernández de Oviedo anteriores a Amador de los Ríos." In *América y la España del Siglo XVI*, edited by Francisco de Solano and Fermín del Pino, 1:117–129. Madrid: Instituto Gonzalo Fernández de Oviedo, 1982.

Cooper, Alix. *Inventing the Indigenous: Local Knowledge and Natural History in Early Modern Europe.* Cambridge: Cambridge University Press, 2007.

Court, William E. "The Ancient Doctrine of Signatures or Similitudes." *Pharmacy in History* 29, no. 3 (September 1999): 41–48.

Cro, Stelio. "La correspondencia epistolar entre el Cardenal Bembo y Fernández de Oviedo: Implicaciones históricas." In *América y la España del Siglo XVI*, edited by Francisco de Solano and Fermin del Pino, 1:53–64. Madrid: Instituto Gonzalo Fernández de Oviedo, 1982.

Cunnally, John. *Images of the Illustrious: The Numismatic Presence in the Renaissance.* Princeton, NJ: Princeton University Press, 1999.

Da Vinci, Leonardo. *Treatise on Painting.* Translated by A. Philip McMahon. Princeton, NJ: Princeton University Press, 1956.

Davies, Martin. "Making Sense of Pliny in the Quattrocento." *Renaissance Studies* 9, no. 2 (1995): 240–257.

DeFelice, Michael. "Prickly Pear Cactus, *Opuntia* spp.: A Spine-Tingling Tale." *Weed Technology* 18, no. 3 (2004): 869–877.

Dickerson, C. D. "Beginnings in Castile." In *Alonso Berruguete: First Sculptor of Renaissance Spain*, edited by C. D. Dickerson III and Mark McDonald, 2–17. Washington, DC: National Gallery of Art, 2019.

———. "Return to Spain, *Pintor del Rey*, and Learning to Sculpt." In *Alonso Berruguete: First Sculptor of Renaissance Spain*, edited by C. D. Dickerson III and Mark McDonald, 36–52. Washington, DC: National Gallery of Art, 2019.

Egaña Rojas, Daniel. "Comerse las Indias: La alimentación como clave clasificatoria del Nuevo Mundo en la obra de Fernández de Oviedo." *Anunario de Estudios Americanos* 72, no. 2 (2015): 579–604.

Espinosa, Aurelio. *The Empire of the Cities: Emperor Charles V, the Comunero Revolt, and the Transformation of the Spanish System*. Boston: Brill, 2009.

Fernández de Oviedo y Valdés, Gonzalo. *Batallas y Quinquagenas*. Edited by Juan Pérez de Tudela Bueso. 4 vols. Madrid: Real Academia de la Historia, [1535–1557] 1983.

———. *Batallas y Quinquagenas*. Edited by Juan Bautista Avalle-Arce. Salamanca: Ediciones de la Diputación de Salamanca, [1535–1557] 1989.

———. *Catálogo real de Castilla*. Edited by Evelia Ana Romano de Thuesen. PhD diss., University of California at Santa Barbara, June 1992.

———. *La historia general de las Indias con privilegio imperial. Fin dela primera parte dela general y natural historia delas indias yslas y tierra firme del mar oceano*. Seville: Juan Cromberger, 1535.

———. *Historia general y natural de las Indias, islas y tierra-firme del mar océano*. Edited by José Amador de los Ríos. 4 vols. Madrid: Imprenta de la Real Academia de la Historia, 1851–1855.

———. *Historia general y natural de las Indias, islas y tierra-firme del mar océano*. Edited by Juan Natalicio González. 14 volumes. Asunción: Editorial Guaranía, 1944–1945.

———. *Historia general y natural de las Indias*. Edited by Juan Pérez de Tudela Bueso. 5 vols. Madrid: Ediciones Atlas, 1959.

———. *Las memorias de Gonzalo Fernández de Oviedo*. Edited by Juan Bautista Avalle-Arce. 2 vols. Chapel Hill: University of North Carolina Press, 1974.

———. *Libro de la cámara real del príncipe don Juan*. Edited by Santiago Fabregat Barrios. Valencia: University of Valencia, [1548] 2006.

———. *Oviedo dela natural hystoría delas Indias*. Toledo: Ramón de Petras, 1526.

———. *Oviedo dela natural hystoría delas Indias*. Toledo: Ramón de Petras, 1526. Facsimile. Chapel Hill: University of North Carolina Press, 1969.

Findlen, Paula, ed. *Empires of Knowledge: Scientific Networks in the Early Modern World*. London: Routledge, 2018.

———. "Natural History." In *The Cambridge History of Science*, edited by Katharine Park and Lorraine Daston, 3:435–468. Cambridge: Cambridge University Press, 2006.

———. *Possessing Nature: Museums, Collecting, and Scientific Culture in Early Modern Italy*. Berkeley: University of California Press, 1994.

———. "The Renaissance of Science." In *The Oxford Illustrated History of the Renaissance*, edited by Gordon Campbell, 378–481. Oxford: Oxford University Press, 2019.

Fisher, John R. *The Economic Aspects of Spanish Imperialism in America, 1492–1810*. Liverpool: Liverpool University Press, 1997.

Fuchs, Leonhart. *De historia stirpium commentarii insignes*. Basel: In officina Isingriniana, 1542. www.huntbotanical.org/library/show.php?10. Accessed May 13, 2022.

Furlotti, Barbara, and Guido Rebecchini. *The Art of Mantua: Power and Patronage in the Renaissance*. Translated by A. Lawrence Jenkens. Los Angeles: J. Paul Getty Museum, 2008.

Gerbi, Antonello. *Nature in the New World: From Christopher Columbus to Gonzalo Fernández de Oviedo*. 1975. Translated by Jeremy Moyle. Pittsburgh, PA: University of Pittsburgh Press, 2010.

Giacomini, Valerio. "L'orto botanico di Napoli." In *Orti botanici delle università italiane*, edited by Mario Bolli, 89–102. Naples: University of Naples, 1965.

Gill, Meredith J. *Augustine in the Italian Renaissance: Art and Philosophy from Petrarch to Michelangelo*. Cambridge: Cambridge University Press, 2005.

Giménez Fernández, Manuel. *Bartolomé de las Casas: Delegado de Ciscernos para la reformación de las Indias (1516–1517)*. Seville: La escuela de estudios hispano-americanos de Sevilla, 1953.

Girolami Cheney, Liana de. "The Impresa in the Italian Renaissance." In *Companion to Emblem Studies*, edited by Peter M. Daly, 251–266. New York: AMS Press, 2008.

Gothein, Marie Luise. "Italy in the Time of the Renaissance and the Baroque Style." In *A History of Garden Art: From the Earliest Times to the Present Day*, edited by Walter P. Wright, 1.207–237. Cambridge: Cambridge University Press, 2014.

Greene, Edward Lee. *Landmarks of Botanical History*. Vol. 2. Redwood City, CA: Stanford University Press, 1983.

Griffin, Clive. *The Crombergers of Seville: The History of a Printing and Merchant Dynasty*. Oxford: Clarendon Press, 1988.

Grubb, James S. *Provincial Families of the Renaissance: Private and Public Life in the Veneto*. Baltimore: Johns Hopkins University Press, 1996.

Harvey, John H. "Gardening Books and Plant Lists of Moorish Spain." *Garden History* 3, no. 2 (Spring 1975): 10–21.

Hernández, Francisco. *The Mexican Treasury: The Writings of Dr. Francisco Hernández*. Edited by Simon Varey. Translated by Rafael Chabrán, Cynthia L. Chamberlin, and Simon Varey. Stanford, CA: Stanford University Press, 2000.

Hortus sanitatis. Moguntaie: Jacobus Meydenbach, 1491. https://catalog.lindahall.org. Accessed November 2, 2022.

Ishikawa, Chiyo. "*La llave de palo*: Isabel la Católica as Patron of Religious Literature and Painting." In *Isabel la Católica, Queen of Castile: Critical Essays*, edited by David A. Boruchoff, 103–119. New York: Palgrave Macmillan, 2003.

Kagan, Richard L. *Clio and the Crown: The Politics of History in Medieval and Early Modern Spain*. Baltimore: Johns Hopkins University Press, 2009.

———. "The Quest for Status and Wealth." In *Alonso Berruguete: First Sculptor of Renaissance Spain*, edited by C. D. Dickerson III and Mark McDonald, 54–62. Washington, DC: National Gallery of Art, 2019.

Keegan, William F. "The 'Classic' Taíno." In *The Oxford Handbook of Caribbean Archaeology*, edited by William F. Keegan, Corinne L. Hofman, and Reniel Rodríguez Ramos, 70–83. Oxford: Oxford University Press, 2013.

Kemp, Martin. *Leonardo da Vinci: The Marvellous Works of Nature and Man*. Oxford: Oxford University Press, [1981] 2006.

———. "Leonardo's Drawings for 'Il cavallo del Duca Francesco di bronzo': The Program of Research." In *Leonardo da Vinci's Sforza Monument Horse: The Art and the Engineering*, edited by Diane Cole Ahl, 64–78. Bethlehem, PA: Lehigh University Press, 1995.

Kent, F. W. *Lorenzo de' Medici and the Art of Magnificence*. Baltimore: Johns Hopkins University Press, 2004.

Kidwell, Carol. *Pontano: Poet and Prime Minister.* London: Duckworth, 1991.

Kohut, Karl. "Fernández de Oviedo, historiador y literato: Humanismo, christianismo, e hidalguía." In *Historia y ficción: Crónicas de América*, coordinated by Ysla Campbell, 43–104. Ciudad Juárez, MX: Universidad Autónoma de Ciudad Juárez, 1992.

Kusukawa, Sachiko. "Leonhart Fuchs on the Importance of Pictures." *Journal of the History of Ideas* 58, no. 3 (1997): 403–427.

Landau, David, and Peter Parshall. *The Renaissance Print, 1470–1550.* New Haven, CT: Yale University Press, 1994.

Landino, Christoforo. *Comento di Christophoro Landino Fiorentino sopra la comedia di Danthe Alighieri poeta Fiorentino.* Bressa: Boninum de Boninis di Raguxi, 1487. https://archive.org/details/McGillLibrary-rbsc_incunabula_Divina-Commedia-di-Dante-Alighieri_1487-20449/page/n3/mode/2up. Accessed October 21, 2022.

Lawrance, J. N. H. "The Spread of Lay Literacy in Late Medieval Castile." *Bulletin of Hispanic Studies* 62, no. 1 (1985): 79–94.

Lillie, Amanda. *Florentine Villas in the Fifteenth Century: An Architectural and Social History.* Cambridge: Cambridge University Press, 2005.

Liss, Peggy K. *Isabel the Queen: Life and Times.* Philadelphia: University of Pennsylvania Press, [1992] 2004.

López Piñero, José María. *Ciencia y técnica en la sociedad española de los siglos XVI y XVII.* Barcelona: Editorial Labor, S.A., 1979.

López Piñero, José María, and María Luz López Terrada. *La influencia española en la introducción en Europa de las plantas americanas: 1493–1623.* Valencia: University of Valencia, 1997.

Luigi, Savastano. *Contributo allo studio critic degli Scrittori agrari italici: Pietro dei Crescenzi.* Acireale: Tip. Orario delle Ferrovie, 1922.

Mallett, Michael, and Christine Shaw. *The Italian Wars, 1494–1559: War, State and Society in Early Modern Europe.* London: Pearson, 2012.

McHam, Sarah Blake. *Pliny and the Artistic Culture of the Italian Renaissance.* New Haven, CT: Yale University Press, 2013.

McIver, Katherine A. *Cooking and Eating in Renaissance Italy: From Kitchen to Table.* Lanham, MD: Rowman and Littlefield, 2015.

Mena García, María del Carmen. *Sevilla y las flotas de Indias: La gran armada de Castilla del Oro (1513–1514).* Seville: Universidad de Sevilla, 1998.

Meneses, Amada. "Andrea Navagero, traductor de Gonzalo Fernández de Oviedo." *Revista de Indias* 18 (1958): 63–72.

Mercadal, J. Garcia, ed. *Viajes de extranjeros por España y Portugal desde los tiempos más remotos hasta fines del siglo XVI.* Madrid: Aguilar, S.A. de ediciones, 1952.

Merrim, Stephanie. "The Apprehension of the New in Nature and Culture: Fernández de Oviedo's *Sumario.*" In *1492–1992: Re/Discovering Colonial Writing*, edited by René Jara, 165–199. Minneapolis, MN: Prisma Institute, 1989.

Milwright, Marcus. "The Balsam of Matariyya: An Exploration of a Medieval Panacea." *Bulletin of the School of Oriental and African Studies, University of London* 66, no. 2 (2003): 193–209.

Mira Caballos, Esteban. "La medicina indígena en la Española y su comercialización (1492–1550)." *Asclepio* 49, no. 2 (1997): 185–198.

Moya Pons, Frank. *La Española en el siglo XVI, 1493–1520: Trabajo, sociedad y política en la economía del Oro.* Santiago, Dominican Republic: Universidad Católica Madre y Maestra Santiago, 1973.

———. "The Establishment of Primary Centres and Primary Plantations." In *General History of the Caribbean*, edited by Pieter C. Emmer, 2:62–78. London: UNESCO, 1999.

———. "The Politics of Forced Indian Labour in La Española, 1493–1520." *Antiquity* 66 (1992): 130–139.

Murray, John. Review of *Chronological History of the West Indies*, by Capt. Thomas Southey. *Quarterly Review* 38 (1828): 193–241.

Myers, Kathleen Ann. *Fernández de Oviedo's Chronicle of America: A New History for a New World*. Austin: University of Texas Press, 2007.

———. "Imitation, Authority, and Revision in Fernández de Oviedo's *Historia general y natural de las Indias*." *Romance Languages Annual* 3 (1991): 523–530.

———. "The Representation of New World Phenomena: Visual Epistemology and Gonzalo Fernández de Oviedo's Illustrations." In *Early Images of the Americas: Transfer and Invention*, edited by Jerry M. Williams and Robert E. Lewis, 183–213. Tucson: University of Arizona Press, 1993.

Nauert, Charles. "Humanists, Scientists, and Pliny: Changing Approaches to a Classical Author." *American Historical Review* 84, no. 1 (1979): 72–85.

Niekrasz, Carmen, and Claudia Swan. "Art." In *The Cambridge History of Science: Early Modern Science*, edited by Katharine Park and Lorraine Daston, 3:773–796. Cambridge: Cambridge University Press, 2006.

Ogilvie, Brian W. "Image and Text in Natural History, 1500–1700." In *The Power of Images in Early Modern Science*, edited by Wolfgang Lefèvre, Jürgen Renn, and Urs Schoepflin, 141–166. Basel: Birkhäuser, 2003.

———. *The Science of Describing: Natural History in Renaissance Europe*. Chicago: University of Chicago Press, 2006.

O'Gorman, Edmundo. *Cuatro historiadores de Indias, siglo XVI: Pedro Martír de Anglería, Gonzalo Fernández de Oviedo y Valdés, Bartolomé de las Casas, Joseph de Acosta*. Mexico City: Secretaría de Educación Pública, 1972.

Orozco, Sebastián de Covarrubias. *Tesoro de la lengua castellana o española*. Madrid: Melchor Sánchez, 1674. Biblioteca virtual Miguel de Cervantes. http://www.cervantesvirtual.com/obra-visor/del-origen-y-principio-de-la-lengua-castellana-o-romance-que-oy-se-vsa-en-espana-compuesto-por-el--0/html/. Accessed October 18, 2020.

Ovejero Pérez, Angela. Introduction to *Libro secondo delle Indie Occidentali*, by Gonzalo Fernández de Oviedo, 7–48. Rome: Bulzoni, 1990.

Paden, Jeremy. "The Iguana and the Barrel of Mud: Memory, Natural History, and Hermeneutics in Oviedo's *Sumario de la natural historia de las Indias*." *Colonial Latin American Review* 16, no. 2 (2007): 203–226.

Panofsky, Erwin. *Renaissance and Renascences in Western Art*. New York: Harper and Row, 1972.

Paz-Sánchez, Manuel de. "'Plantain of Guinea': The Atlantic Adventure of Banana." *Global Journal of Science Frontier Research* 14, no. 2 (2014): 49–65.

Pemán y Pemartín, César. *Juan van Eyck y España*. Cadiz: Museo Provincial de Bellas Artes de Cadiz, 1969.

Petrarch. *Petrarch's Remedies for Fortune Fair and Foul: Book 1*. 1354–1366. Translated by Conrad H. Rawski. Bloomington: Indiana University Press, 1991.

———. *The Sonnets, Triumphs, and Other Poems of Petrarch*. London: George Bell and Sons, 1879.

Pliny the Elder. *Naturalis historia*. Translated by John Bostock. London: Taylor and Francis, [77–79 AD] 1855. http://data.perseus.org/texts/urn:cts:latinLit:phi0978.phi001. Accessed October 21, 2022.

Pontano, Giovanni. *I trattati delle virtù sociali: De Liberalitate, De Beneficentia, De Magnificentia, De Splendore, De Conviventia*. Rome: Edizioni dell'Anteneo, [1498] 1965.

Portuondo, María M. "Cosmography at the *Casa*, *Consejo*, and *Corte* During the Century of Discovery." In *Science in the Spanish and Portuguese Empires, 1500–1800*, edited by Daniela Bleichmar, Paula De Vos, Kristine Huffine, and Kevin Sheehan, 57–77. Stanford, CA: Stanford University Press, 2009.

Prieto, Andrés I. "Classification, Memory, and Subjectivity in Gonzalo Fernández de Oviedo's *Sumario de la natural historia* (1526)." *MLN* 124, no. 2 (2009): 329–349.

———. *Missionary Scientists: Jesuit Science in Spanish South America, 1570–1810*. Nashville, TN: Vanderbilt University Press, 2011.

Rabasa, José. *De la invención de América: La historiografía española y la formación del eurocentrismo*. Translated by Aldo Mazzucchelli. Mexico City: Universidad Iberoamericana, [1993] 2009.

Rackham, H. Introduction to *Natural History*, by Pliny the Elder. Translated by H. Rackham, vii–xiv. Cambridge, MA: Harvard University Press, 1938.

Ramos Pérez, Demetrio. "Fernández de Oviedo y el 'enigma' de la edición de 1547 de su *Historia general*." *Revista historia* 1, no. 8 (1969): 443–461.

———. "Las ideas de Fernández de Oviedo sobre la técnica de colonización en América." *Cuadernos hispanoamericanos* 17 (1957): 279–289.

Real Academia Española. *Diccionario de la lengua española*. https://dle.rae.es/.

———. *Diccionario panhispánico del español jurídico*. https://dpej.rae.es/.

Reeds, Karen M. "Leonardo da Vinci and Botanical Illustration: Nature Prints, Drawings, and Woodcuts ca. 1500." In *Visualizing Medieval Medicine and Natural History, 1200–1550*, edited by Jean A. Givens, Karen M. Reeds, and Alain Touwaide, 205–237. Aldershot: Ashgate, 2006.

Ríos, José Amador de los. "Vida y escritos de Gonzalo Fernández de Oviedo y Valdés." In *Historia general y natural de las Indias*, by Gonzalo Fernández de Oviedo, edited by José Amador de los Ríos, 9–107. Madrid: Real Academia de la Historia, 1851.

Rodríguez Morel, Geraro, ed. *Cartas de la real audiencia de Santo Domingo (1530–1546)*. Santo Domingo: Archivo general de la nación, 2007.

Romano de Thuesen, Evelia Ana. Abstract to *Catálogo real de Castilla*, by Gonzalo Fernández de Oviedo, viii–ix. Edited by Evelia Ana Romano de Thuesen. PhD diss., University of California at Santa Barbara, June 1992.

Ruggles, Fairchild D. *Gardens, Landscape, and Vision in the Palaces of Islamic Spain*. University Park: Pennsylvania State University Press, 2000.

Salas, Alberto Mario. *Tres cronistas de Indias: Pedro Mártir de Anglería, Gonzalo Fernández de Oviedo, Fray Bartolomé de las Casas*. Mexico City: Fondo de Cultura Económica, 1986.

Salas, Ximenez de. "Jaime Serrat, pintor del Príncipe Don Juan." *Archivo español de arte y arqueología* 12, no. 36 (1936): 269–272.

Sánchez Jiménez, Antonio. "Memoria y utilidad en el *Sumario de la natural historia de las Indias* de Gonzalo Fernández de Oviedo." *Colonial Latin American Review* 13, no. 2 (2004): 263–273.

Sauer, Carl Ortwin. *The Early Spanish Main*. Berkeley: University of California Press, 1969.

Saunders, Gill. *Picturing Plants: An Analytical History of Botanical Illustration*. Berkeley: University of California Press, 1995.

Schäfer, Ernst. "Antonio de Villasante, descubridor droguista en la Isla Española." *Investigación y proceso* 9, no. 1 (1935): 13–15.

Scher, Stephen K. "An Introduction to the Renaissance Portrait Medal." In *Perspectives on the Renaissance Medal*, edited by Stephen K. Scher, 1–9. New York: Garland, 2000.

Seville, Isidore of. *The Etymologies of Isidore of Seville*. Edited by Stephen A. Barney. Cambridge: Cambridge University Press, 2006.

Shestack, Alan. Introduction to *Complete Engravings of Martin Schongauer*, v–xvi. New York: Dover, 1969.

Smith, Pamela H. *The Body of the Artisan: Art and Experience in the Scientific Revolution*. Chicago: University of Chicago Press, 2004.

Stannard, Jerry. *Herbs and Herbalism in the Middle Ages and Renaissance*. Edited by Katherine E. Stannard and Richard Kay. Farnham: Ashgate, 1999.

Sturtevant, William. "The Sources for European Imagery of Native Americans." In *New World of Wonders: European Images of the Americas, 1492–1700*, 24–33. Seattle: University of Washington Press, 1992.

Tagliolini, Alessandro. *Storia del giardino italiano: Gli artisti, l'invenzione, le forme dall' antichità al XIX secolo*. Florence: La casa Usher, 1988.

Teglia, Vanina. "Claroscuros del archive colonial: La escritura sobre la naturaleza de Fernández de Oviedo." *Huarte de San Juan: Geografía e Historia* 27 (2020): 267–290.

———. "Conocer el Nuevo Mundo: Referencialidad en las ilustraciones de Fernández de Oviedo." *Olivar* 15, no. 2 (December 2014): 1–14.

Tito Rojo, José. "Permanencia y cambio en los jardines de la Granada morisca (1492–1571): Los jardines de los palacios nazaríes; La Alambra y el Generalife." In *Jardín y naturaleza en el reinado de Felipe II*, edited by Carmen Añón and José Luis Sancho, 363–379. Madrid: Sociedad estatal para la conmemoración de los centenarios de Felipe II y Carlos V, 1998.

Tito Rojo, José, and Manuel Casares Porcel. "From the Andalusí Garden to the Andalusian Garden: Remnants and Re-Creation." In *Middle East Garden Traditions: Unity and Diversity*, edited by Michel Conan, 287–305. Washington, DC: Dumbarton Oaks Research Library and Collection, 2007.

Tudela Bueso, Juan Pérez de. "Vida y escritos de Gonzalo Fernández de Oviedo." In *Historia general y natural de las Indias*, by Gonzalo Fernández de Oviedo, 1:7–169. Madrid: Ediciones Atlas, 1959.

Turner, Daymond. "The Aborted First Printing of the Second Part of Oviedo's *General and Natural History of the Indies*." *Huntington Library Quarterly* 46, no. 2 (1983): 105–125.

———. "Forgotten Treasure from the Indies: The Illustrations and Drawings of Fernández de Oviedo." *Huntington Library Quarterly* 48, no. 1 (1985): 1–46.

———. "Gonzalo Fernández de Oviedo's *Historia General y Natural*: First American Encyclopedia." *Journal of Inter-American Studies* 6, no. 2 (April 1964): 267–274.

———. "Los libros de alcaide: La biblioteca de Gonzalo Fernández de Oviedo y Valdés." *Revista de Indias* 31 (1971): 109–198.

Villasante, Antonio de. "Unnamed." In *Santo Domingo en los manuscritos de Juan Bautista Muñoz*, edited by Roberto Marte, 449–453. Santo Domingo: Ediciones Fundación García Arévalo, 1981.

Vives, Juan Luis. *Vives: On Education*. With introduction by Foster Watson. Lanham, MD: Rowman and Littlefield, [1531] 1971.

Watts, David. *The West Indies: Patterns of Development, Culture and Environmental Change Since 1492*. Cambridge: Cambridge University Press, [1987] 1994.

Weiss, Jessica Sue. "Isabel of Castile, Flemish Aesthetics, and Identity Construction in the Works of Juan de Flandes." PhD diss., University of Texas at Austin, 2014.

———. "Juan de Flandes and His Financial Success in Castile." *Journal of Historians of Netherlandish Art* 11, no. 1 (Winter 2019): 1–38.

Weniger, Del. *Cacti of the Southwest: Texas, New Mexico, Oklahoma, Arkansas, and Louisiana*. Austin: University of Texas Press, 1969.

Wey Gómez, Nicolás. "Memorias de la zona tórrida: El naturalism clásico y la 'tropicalidad' Americana en el *Sumario del natural historia de las Indias* de Gonzalo Fernández de Oviedo (1526)." *Revista de Indias* 259 (2013): 609–632.

———. *The Tropics of Empire: Why Columbus Sailed South to the Indies*. Cambridge, MA: MIT Press, 2008.

Williams, Gareth D. *Pietro Bembo on Etna: The Ascent of a Venetian Humanist*. Oxford: Oxford University Press, 2017.

Yarza Luaces, Joaquin. *Los reyes católicos: Paisaje artístico de una monarquía*. Madrid: Editorial Nerea, 1993.

INDEX

ACKNOWLEDGMENTS

It is my privilege to be able to thank those who have made the writing of this book an experience of great learning and much joy. Neither pen nor paintbrush would be sufficient to convey my appreciation for the support and guidance of Rolena Adorno. In addition to setting me on this journey, she has shown me the makings of a true scholar, mentor, and teacher. I have learned much from her example and counsel, and I am incredibly grateful for the enormous amount of time and energy she has invested in me over the years. I will always cherish the memories of our time together.

For his careful reading and thoughtful comments on (multiple) drafts of chapters, my heartfelt thanks to David Stark. His editorial skills are matched only by his steadfast support for me. I am fortunate to have such an advocate. Others also offered much-appreciated suggestions on various chapters; my gratitude to Médar Serrata, Philip Gansen, and Megan Eding for having taken a deep jump into unfamiliar territory. I thank David Crane for fielding my questions regarding the classical world and for talking out sticky grammar points with me.

Much of this book was researched and written during my first years at Grand Valley State University in Allendale, Michigan. I could not have hoped for a more welcoming department than that of Modern Languages and Literatures, nor for more supportive colleagues and friends who have made Grand Rapids feel like home (some of whom have since moved away). I owe a special thanks to Yasmin Gopal for her friendship and hospitality, as the gracious host of countless dinners and work sessions. A stroke of luck landed me in an early modern writing group organized by the Newberry Library during the first years of COVID-19 with Tatevik Gyulamiryan and Molly Borowitz, a welcome and much-needed reprieve from the solitude of those months and a source of inspiration and learning. David Barrons has been an invaluable interlocutor with whom I have shared many an idea, and who has been a buttress against cloudy days.

This book is indebted to the institutional support of Grand Valley State University, and specifically to several research grants I was awarded by the Center for Scholarly and Creative Development in 2018 and 2020 as well as its Book Subvention Fund. A big thanks to the wonderful Document Delivery team at the GVSU library, and in particular to Amy Duwe, without whose help this book would have been much more difficult to write. A fellowship from the Huntington Library in July 2018 allowed me to study many of the extant natural history books from the *Historia general* in Oviedo's hand amid its beautiful botanical gardens. Chapter 7 would not have been possible were it not for the generosity and help of William Reese (1955–2018), Rolena Adorno, and the manuscript's current owner, who gave me permission to reproduce the photographs that I took of its pages.

My experience working with Robert Lockhart at the University of Pennsylvania Press has exceeded all my expectations. I thank him for taking an interest in my project and for his feedback and support. I am grateful too for the suggestions of Peter C. Mancall and the two anonymous readers of the manuscript of *Natural Designs*. Finally, I thank all those at the University of Pennsylvania Press who had a hand in making this book a reality, and whose professionalism and hard work has made my experience as a first-time author such a positive one.

* * *

I dedicate this book to my parents, for all they have done for me, and to my sister, for always being in my corner.

CPSIA information can be obtained
at www.ICGtesting.com
Printed in the USA
JSHW080327200723
44547JS00002B/2

9 781512 824568